Reviews

The question whether mathematics is discovered or created divides the mathematical community into two camps. Some mathematicians—like me—are in one camp in the morning and in the other one in the afternoon. My opinion, I confess, depends on the type of work being done at the moment. But deep in my mind, I am fully convinced that, based on some very elementary and not yet understood endowment of our brain, the fantastic mathematical universe is human-made. This can't be proved mathematically. The best one can hope for are compelling arguments and strong empirical evidence.

This is what Klaus Truemper's book "The Construction of Mathematics: The Human Mind's Greatest Achievement" delivers. It sheds surprising and fascinating new light on the issue. Powerful arguments are provided by using the method of language games invented by the philosopher Ludwig Wittgenstein. Employing results of modern brain science about human cognition, the book also explains how it is possible that eminent mathematicians and scientists arrive at diametrically opposed answers for the creation vs. discovery question. Truemper's findings are consistent with my ultimate conviction.

> —**Martin Grötschel, mathematician and President, Berlin-Brandenburg Academy of Sciences and Humanities, Germany**

Klaus Truemper has made an original and daring attack on the foundations of mathematics. Readers will enjoy his forthright and unswerving analysis. His ideas should become recognized and influential.

> —**Reuben Hersh, mathematician and award-winning author of a number of books on the nature, practice, and social impact of mathematics**

As computational methods become increasingly powerful, the engineer of today often resorts to simulation methods and is acutely aware of the limitations of contrived analytical mathematical methods. Truemper's exposition puts into focus the debate as to whether mathematics is really intrinsic to the physical world or is in fact made up as we go along. The elucidation the book delivers on this topic is of significance, as science often moves along faster when we are released from anachronistic notions that imprison freedom of thought. Klaus Truemper permits us to be more daring with our mathematics.

—Derek Abbott, physicist and engineer, University of Adelaide, Australia

Is mathematics the product of human creativity and ingenuity or does it exist independently of mankind and only waits to be discovered? Is the latter the reason for its incredible success or is mathematics not so indispensable in our world after all?

After a fascinating tour through the history of mathematics and computation that provides astonishing new perspectives, Klaus Truemper's book addresses the philosophical question of creation versus discovery from many different directions ranging from Wittgenstein's philosophy to brain science. Pros and cons are collected and discussed in a comprehensive and—given the difficult subject— remarkably light and entertaining way and: the book comes to a conclusion.

Is the issue finally settled? No! Can it possibly be? Most certainly not! The given circumstantial evidence does, however, stimulate the reader to rethink his or her point of view, question his or her own arguments, rethink the given reasoning and sharpen the issue. The book is a pleasure to read!

—Peter Gritzmann, mathematician, author of mathematics and popular science books

This wonderful book addresses the oldest and thorniest question in the philosophy of mathematics: Is mathematics discovery or invention? The key contribution of this book is to use Ludwig Wittgenstein's technique of language games to shed light on this deep philosophical question. Overall this is the most insightful and compelling contribution to the debate that I have read and I am inclined to agree with the conclusions.

But the book is much more than an important contribution to a fundamental debate. It is beautifully and lucidly written and, in contrast to most texts on philosophical matters, is enjoyable and easy to read. Mathematicians know that a picture is worth a thousand words, and the rich use of beautifully presented illustrations genuinely enriches the readers experience.

The book begins with a short history of the major themes of mathematics. It is, by a country mile, the most insightful such history that I have read. An intelligent reader with no interest whatsoever in philosophy would still learn much from this book. I cannot recommend it too highly.

—Geoffrey Whittle, mathematician, Victoria University of Wellington, New Zealand

Klaus Truemper's "The Construction of Mathematics: The Human Mind's Greatest Achievement" is a unique blend of history and penetrating philosophical analysis. After taking the reader on a dizzying journey through the history of mathematics and computation, Truemper arrives at his central question: Is mathematics discovered or invented?

Relying on a range of philosophical approaches and some brilliant argumentation, he arrives at what seems like the only possible answer. Mathematics, he shows, is not "out there," waiting to be discovered; it is, rather, the highest creation of the human mind.

Truemper's book is not only insightful and original but also fast-paced and gripping. Whether you are a mathematician, historian, philosopher, or layman, you will find it thought-provoking as well as highly enjoyable.

—Amir Alexander, historian of science and award-winning author of books on the interconnection of mathematics and its social, cultural, and political settings

Also by Klaus Truemper

Brain Science

Artificial Intelligence
Wittgenstein and Brain Science
Magic, Error, and Terror

History

The Daring Invention of Logarithm Tables

Technical

Logic-based Intelligent Systems
Effective Logic Computation
Matroid Theory

Edited by Ingrid and Klaus Truemper

F. Hülster *Introduction to Wittgenstein's*
Tractatus Logico-Philosophicus
(English and German edition)

F. Hülster *Berlin 1945: Surviving the Collapse*

THE CONSTRUCTION
OF MATHEMATICS

THE HUMAN MIND'S
GREATEST ACHIEVEMENT

KLAUS TRUEMPER

Leibniz Company

Softcover published by Leibniz Company
2304 Cliffside Drive
Plano, Texas, 75023
USA

Original edition 2017
Updated editions 2018, 2019, 2021

Cover Art:

"Mathematics of Mankind" by Andrew Ostrovsky copyright © 2017 used under exclusive license. Cover design by Ingrid Truemper.

The book is typeset in LATEX using the Tufte-style book class, which was inspired by the work of Edward R. Tufte and Richard Feynman.

Sources and licensing arrangements for all figures are included in the Notes section.

Library of Congress Cataloging-in-Publication Data
Truemper, Klaus, 1942–

The Construction of Mathematics: The Human Mind's Greatest Achievement
Includes bibliographical references and subject index.
ISBN 978-0-9663554-3-7
1. Mathematics. 2. Construction

Contents

1

Introduction

Over the span of tens of thousands of years, humans have created an elaborate body of theory unequaled in size and complexity: mathematics. Early work on that edifice was slow, with the pace of progress measured in thousands of years. But gradually things speeded up; progress was achieved in hundreds of years, then in decades. And now the theory is leaping forward at a dizzying pace.

There are hundreds of books chronicling that development either in multi-volume works attempting to cover the entire spectrum of results, or in more specialized texts investigating specific areas or periods.[1] Amazingly, some authors—for example Florian Cajori (1859–1930) and James Roy Newman (1907–1966)—accomplished both goals.[2]

This book is different. It poses questions, evaluates answers, and develops conclusions about mathematics as a creation of the human mind.

Chapters 2–7 provide a longitudinal overview of important parts of mathematics that, by their very nature, can be covered in everyday language and appealing simplicity.

The key idea is to bring out the struggle for insight in the face of very difficult questions; the paths of investigation that often proved to be correct, but sometimes to be dead ends; and the triumphant clarity eventually achieved.

There is a profound philosophical question: Where do all the results of this gigantic body of theory come from? Are they already present in some hidden, possibly metaphysical, location and then discovered by inquisitive minds, or are they created in the same way that engineers design various machines for energy conversion, production of goods, or transportation?

Chapter 8 reviews this question and prior answers. They claim either creation, or discovery, or some combination of both. Obviously, many of these answers are in conflict. Yet each of them is defended with seemingly irrefutable arguments.[3] In the second part of the book, we investigate this question and work toward a resolution.

Chapters 9 and 10 introduce and then use the concept of language games first proposed by the philosopher Ludwig Wittgenstein (1889–1951) for the resolution of philosophical problems. A *language game* is a controlled setting of language use that brings a particular facet of a given philosophical problem into focus. One then imagines the operation of the language game and thus gains insight into that facet.

For the identification of relevant facets and the subsequent construction of the corresponding language games, one initially assembles a large number of example situations involving the given philosophical problem. For the case at hand, Chapters 2–6 supply many example situations. From these, we identify a number of facets, then build and operate a language game for each of them.

In some of the language games, mathematical developments are linked with creative processes of the arts, in particular sculpture and music composition.

Other language games involve imagined events, processes, or dialogues. During the operation of each language game, we always assume that mathematics is discovered. It turns out that the operation of each language game then produces some contradictory conclusion, indicating that the assumption of discovery is flawed.

Chapters 11 and 12 look at particular arguments in favor of the discovery claim. A strong argument for discovery is based on the seemingly astonishing agreement of mathematics and nature. For example, Einstein's mathematical theory of relativity explains how space and time of nature are linked. Such universal agreement is only possible if mathematics is part of nature—so the argument goes—and thus mathematics can only be discovered.

We examine that argument and see that a different interpretation is possible: We have *shaped* mathematics so that we can represent processes of nature. Moreover, we have *defined* the processes supposedly occurring in nature. It's no wonder that *our* mathematics can represent *our* processes!

There also are many more stories of failed mathematical models versus successful ones, and we have a tendency to ignore, gloss over, or simply forget those failures.

Then there are processes of nature—of course, as we have defined them—where we have not been able to build any mathematical model or carry out certain computations. For those cases, mathematical theories have been built that rule out such model building or computation. How could these theories be part of nature?

Yet another aspect is the following. There are mathematical theories that are in blatant violation of the processes of nature as we have defined them. How can these theories be part of nature?

Another argument in favor of discovery is that humans simply cannot live without mathematics, and thus that mathematics must be part of nature. For a counterargument to this claim, we look at a community of humans that live successfully and happily without any mathematics at all.

For more than 30 years, we have repeatedly posed the question of creation versus discovery of mathematics to mathematicians. Amazingly, the vast majority voted for discovery. We say "amazingly," since opinions of eminent mathematicians of the 19th and early 20th century were mixed. For example, Georg Cantor (1845–

1918), Richard Dedekind (1831–1916), and Carl Friedrich Gauss (1777–1855) argued for creation, while Gottlob Frege (1848–1925) and Kurt Gödel (1906–1978) claimed that discovery takes place. How can these differences be explained, in particular the shift toward discovery?

For an answer, Chapter 13 takes a critical look at the brain, which after all is the machinery that makes and evaluates the arguments swirling around the question.

Brain science has made incredible strides since the 1990s. In this book we use the new results and investigate how the structure of the brain may impact our answers. In the process, we formulate a conjecture that sheds light on the reason for the different opinions.

Chapter 14 summarizes the various arguments. In the face of so many contradictions arising from the discovery hypothesis, we come to the conclusion that mathematics is indeed created by the human mind. That conclusion fits the historical facts about mathematical developments, does not require metaphysical concepts, and is consistent with the view that the human activities of music composition, sculpture, and writing are creation. The conclusion is also consistent with the notion of Occam's razor[4] of science, according to which the simpler explanation is preferred when there is a choice.

Above all, reading the book is meant to be an enjoyable journey across a wide-ranging territory of human thought and accomplishment. If you agree in the end that this goal has been achieved, we have done our job.

Technical Remarks

Chapters 2–14 don't require any mathematical background beyond everyday knowledge of numbers and the elementary operations of addition, subtraction, multiplication, and division. The subsequent, extensive, Notes section expands upon the discussion and justifies various statements using at most high school mathematics. The Notes section also points to references and material readily available on the Internet, and it lists the licenses under which the various figures are included in the book. The reader may choose to ignore the latter notes since they are only of legal interest.

Portraits are included for virtually all mathematicians mentioned in the book, courtesy of Wikipedia and various academic and research institutions. The portrait appears on first mention of each mathematician.

2

Numbers

We all know the *whole numbers*, which consist of 1, 2, 3, ..., their negative counterparts −1, −2, −3, ..., and the 0. We also know the *fractions* such as $\frac{1}{2}$ or $\frac{3}{5}$, where a whole number is divided by a nonzero whole number.

These numbers were developed over tens of thousands of years for commerce and trade. Today's computers still use those numbers, and in some sense, these are the only numbers computers know and are able to process.

But for mathematicians, the idea of whole numbers and fractions, useful as they are, did not suffice. So over the past 4,000 years, they created a variety of additional numbers.

At the core of this creative process are the *decimal numbers*. Examples are 0.34714 ... or 7.31739 ..., where the digits may stop at some point or continue indefinitely. How many more numbers were so created? The astounding result is: Many, many more beyond the whole numbers and fractions. Let's try to understand this with an analogy.

Abundance of Decimal Numbers

Consider the oceans of the earth. They contain more than 300 million cubic miles of water.[5] Let this huge volume of water represent the number of decimal numbers. Which portion of that volume corresponds to the whole numbers and fractions?

Until the mid-19th century, this question wasn't posed let alone answered, due to the simplistic notion that there were an infinite number of whole numbers, of fractions, and of decimal numbers, and that one could not say more about these quantities. But then Georg Cantor (1845–1918) started a revolution by showing that there are various types of infinity.

Georg Cantor.[6]

Cantor's results imply that the water representing the whole numbers and fractions, among the vast amount of water representing the decimal numbers, constitutes less than a drop of water! In fact, it is less than one molecule of water. Indeed, it can only be said there is *no* smallest amount of water that would represent the whole numbers and fractions!

Thus, mathematicians have created an incredible number of numbers beyond those needed for commerce and trade. But they have also created an astounding *variety* of numbers. The rest of this chapter covers that development.

To start, we need to introduce standard mathematical terminology. Whole numbers are called *integers*, fractions are called *rational numbers* since they are ratios of whole numbers, and *decimal numbers* are a particular representation of the *real numbers*. From now on, we will use that standard terminology. We begin with the integers.

Integers

Counting began at least 50,000 years ago.[7] We will skip a detailed discussion of the innovative ways in which counts were defined, for example, by *body parts*.[8] Around 10,000 BCE, herder-farmers invented sophisticated pebble counting where, for example, 10 pebbles were replaced by one specially marked pebble.[9]

As commerce and trade developed, so did the representation of numbers, computational processes, and methods for recording results. We defer discussion of that development until Chapter 3, and instead cover here the evolution of numbers using the decimal notation of modern mathematics.

We start out with the positive integers 1, 2, 3, ... They usually are called the *natural numbers*. A seemingly trivial observation was that a natural number may be produced by several different multiplication steps. For example, $2 \cdot 12$ is equal to 24, as is $3 \cdot 8$. This fact has little use in commerce, but the ancient Greeks thought it worthy of investigation.

The first observation was that some natural numbers n are not the result of any multiplication step except for the trivial $1 \cdot n$. Examples are $n = 2, 3$, and 5. Such numbers are now called *primes*.

The second observation was that multiplication just using prime numbers can create any natural number. For example, in the multiplication step $2 \cdot 12 = 24$, the *factor* 12 is the result of the multiplication step $2 \cdot 6 = 12$. Hence, we can write $2 \cdot 2 \cdot 6 = 24$. Continuing in this fashion, we can replace each nonprime factor by the product of two smaller numbers, until we have a final representation where all factors are primes. Here, we get $2 \cdot 2 \cdot 2 \cdot 3 = 24$.

These facts piqued the curiosity of the ancient Greeks. In particular, Euclid (mid 4th–mid 3th century BCE) posed the following two questions.

First, are there an infinite number of primes? Euclid proved that this is so.[10]

Second, is the representation of a natural number by prime factors unique except for trivial reordering of the factors? Euclid also proved this to be the case.[11] This result is now known as the *Fundamental Theorem of Arithmetic*.[12]

The primes occur within the natural numbers in a seemingly irregular pattern. Mathematicians since Euclid have striven for a predictive formula, and in the process have created an astonishing wealth of results. They are brought together in the book *The Music of the Primes*.[13]

Division of an integer quantity by another integer quantity, an important operation for commerce, forced creation of the rational numbers; examples are $\frac{1}{4}$ and $\frac{2}{3}$. We look at these numbers next.

Euclid, by Justus van Gent, ca. 1474.[14]

Rational Numbers

In ancient times, it was believed that all measurements in nature could be expressed with rational numbers, using a suitable unit length for each measurement. That belief postulated a unity of nature and mathematics that simply does not exist. The decisive blow came by a member of the Pythagorean society during the 5th century BCE, possibly Hippasus of Metapontum.[15] He showed that the length of the diagonal of a square with length 1 for each side, which was known[16] to be $\sqrt{2}$, is not a rational number.[17]

We should mention that the ancient Babylonians already knew the length of the diagonal to be $\sqrt{2}$ more than 1,000 years before Pythagoras. Indeed, they computed the value of $\sqrt{2}$ with high precision, as proved by clay tablet YBC 7289, created around 1800–

Clay tablet YBC 7289, ca. 1800–1600 BCE. Size about 8cm each side.[18]

1600 BCE. It displays the square and shows on the diagonal for $\sqrt{2}$ a Babylonian number that in decimal notation is 1.41421297 and thus correct for five digits after the decimal point. The Babylonian number below the diagonal approximates $\frac{1}{\sqrt{2}}$.

A number that is not rational is called *irrational*. Thus, $\sqrt{2}$ is an irrational number. We look at the irrational numbers next.

Irrational Numbers

How can we tell rational and irrational numbers apart? For a simple rule, we need to look at the conversion of rational numbers to decimal numbers. The conversion is also called *decimal expansion*. For example, $\frac{1}{4}$ becomes 0.25, and $\frac{3}{4}$ becomes 0.75. The decimal expansion need not have finite length. For example, $\frac{2}{3}$ becomes 0.666 . . . and thus involves an infinite sequence of digits.

In each infinite case, the expansion has a particular structure. That is, from some point on in the decimal expansion, a string is repeated over and over. For example, $\frac{628}{2475}$ has the decimal expansion 0.25373737 . . . . Thus, the expansion begins with 0.25, then repeats the string 37 indefinitely.

The irrational numbers also have infinite decimal representation, but not with the repetitive structure exhibited by the rational numbers. This is the decisive difference.

Real Numbers

The *real numbers* are the collection of rational and irrational numbers. An immediate question is: Can the real numbers be systematically constructed? The rational numbers are easily handled by systematic enumeration of ratios of integers. But how can the irrational numbers be constructed? Naïvely, we could think of creating an irrational number as follows: We write down a finite number of digits, then enter the decimal point, and from then on keep on writing digits indefinitely while avoiding any repetitive pattern. The description is technically correct, but gives no insight how the irrational numbers are distributed between the rational ones. Richard Dedekind (1831–1916) described an ingenious construction of the irrational numbers from the rational ones that provides that insight. The construction is now called the *Dedekind cut*. It can be summarized as follows.

Richard Dedekind.[19]

Sort the rational numbers, then cut the sorted list into two lists. Under a certain condition, an irrational number lies in the cut.[20] When this process is repeated for all cuts, all irrational numbers are produced.

Algebraic Numbers

The Dedekind cut constructs all irrational numbers from the rational ones. But so far we have seen just one explicit example of an irrational number, $\sqrt{2}$. How can we construct more irrational numbers? The example $\sqrt{2}$ gave mathematicians an idea: $\sqrt{2}$ is the solution of the equation $x^2 = 2$, which is the same as $x^2 - 2 = 0$. So they defined equations with a single variable x where the right-hand side is 0, and where the left-hand side is a sum of terms $a_k \cdot x^k$ where $k \geq 0$ and a_k are integers. The left-hand side is called a *polynomial with integer coefficients*.[21] Example polynomials are $3x^2 + 15x - 27$ and $36x^5 - 17x^2 - 19$. A solution of the equation is called a *root* of the polynomial.

A number is *algebraic* if it is the root of some polynomial with integer coefficients. The construction of the algebraic numbers has the nice feature that the integers and the rational numbers[22] are algebraic. But there are also lots of examples where a root is irrational. For example, for any integer $k \geq 2$ and any positive integer n, the polynomial $x^k - n$ produces irrational roots if n is not equal to some integer raised to the kth power.

Given the abundance of irrational numbers that are algebraic, the question immediately arose: Are *all* irrational numbers actually algebraic and thus roots of polynomials with integer coefficients?

The answer was long in coming. Leonhard Euler (1707–1783) defined the name *transcendental* for irrational non-algebraic numbers, but evidently did not provide an example.[25]

Joseph Liouville (1809–1882) first proved the existence of transcendental numbers[26] in 1844. So, yes, there are irrational numbers that are not algebraic.

Leonhard Euler, by Jakob Emanuel Handmann, 1756.[23]

Joseph Liouville.[24]

Imaginary Numbers

The computation of roots of polynomials goes back to ancient times. We will skip discussion of the varied history and mention only that time and again a pesky problem surfaced: A root could not be computed because, for some positive number a, a term of the form $\sqrt{-a}$ occurred. Indeed, there is no real number such that multiplication with itself produces a negative number, and thus it was argued that a number $\sqrt{-a}$ did not exist. Eventually, the definition $i = \sqrt{-1}$ was used to reformulate $\sqrt{-a}$ as $\sqrt{a} \cdot i$. The numbers involving i are called *imaginary*. The motivation for the term was that imaginary numbers, though required for computation, were not considered to be actual numbers. This view persisted up to the time of Euler, who established fundamental results involving imaginary numbers.[27]

Complex Numbers

For real numbers a and b, the term $a + bi$ is defined to be a *complex* number. It can be plotted in the *complex plane*[28] where the x-axis is real and the y-axis is imaginary. In that representation, a complex number can be viewed as an arrow, or technically a *vector*, going from the origin of the plane to the plotted point. Operations with complex numbers then become intuitive steps involving vectors.[29]

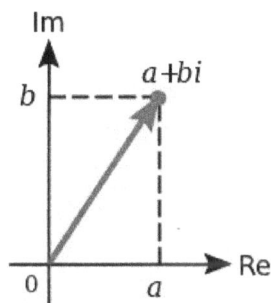

Representation of the complex number $a + bi$ in the complex plane, with real axis Re and imaginary axis Im.[30]

Two Special Numbers: e and π

In subsequent chapters, we repeatedly meet two special numbers: $e = 2.7182\ldots$, called the *Euler number*, and $\pi = 3.1415\ldots$. The number e can be defined in various ways.[31]

For example, $e = \sum_{n=0}^{\infty} \frac{1}{n!}$ where $n! = 1 \cdot 2 \cdot 3 \cdot \ldots \cdot n$.

Jacob Bernoulli (1655–1705) established e while investigating the effect of compounding interest.[33] The number π relates the diameter d of a circle to its circumference c by $c = d \cdot \pi$.

Jacob Bernoulli, by Niklaus Bernoulli (1662-1716).[32]

The numbers e and π are not rational; indeed they are transcendental.[34] Charles Hermite (1822–1901) proved this result for e in 1873, and Ferdinand von Lindemann (1852–1939) for π in 1882 using Hermite's method.

The Creation of Numbers

We have seen a variety of numbers. Where do they come from? Let's begin with the natural numbers. A struggle over tens of thousands of years[37] created the natural numbers and mathematical operations for their use in commerce and trade. That it took such a sustained effort implies that there is nothing obvious about the natural numbers, their representation, or their use. This conclusion in principle does not invalidate metaphysical arguments that, somehow and somewhere outside nature, the numbers and indeed all results of mathematics are stored and ready to be retrieved by the human mind as needed.

Charles Hermite, ca. 1901.[35]

Ferdinand von Lindemann.[36]

Later in this book, we shall address such arguments. For the moment, we only mention the metaphysical claim by Leopold Kronecker (1823–1891) that "God made the integers, all else is the work of man."[38]

The claim was quite appropriate for Kronecker, since he believed in *finitism* of mathematics, where infinite collections are ruled out.[39] But the collection of natural numbers is infinite! Kronecker resolved the conflict between that fact and finitism by appealing to a metaphysical construction.

In the 19th century, it was recognized that the natural numbers, familiar as they were, had not been concisely defined. For discussion of a suitable definition, we need the concept of a *set*, which for our purposes is a collection of *elements*. For example, a set S with three elements a, b, and c is denoted by $S = \{a, b, c\}$. The set of natural numbers, say N, is then $N = \{1, 2, 3, \ldots\}$. We also need the concept of *axioms*, which are statements postulating the existence of mathematical items and relationships among them. Thus, out of thin air, just using the imagination of the human mind, these items and relationships come into existence.

Leopold Kronecker.[40]

Several constructions of N have been proposed.[42] The axioms by Giuseppe Peano (1858-1932), now called *Peano axioms*, are now universally used.[43]

From the set N, we readily define the integers by introducing negation, and the rational numbers by ratios of integers. Next, the Dedekind cut creates the irrational numbers from the rational numbers. The cut is an axiom, as is clearly stated by Dedekind.[44]

Giuseppe Peano.[41]

Then the algebraic numbers are defined from the integers using roots of polynomials, and the transcendental numbers are all real numbers that are not algebraic.

Finally, we obtain the imaginary and complex numbers from the real numbers by defining $i = \sqrt{-1}$ and, for any real a and b, the number $a \cdot i$ to be imaginary, and the number $a + b \cdot i$ to be complex.

Summary

The integers together with the rational and irrational numbers constitute the real numbers. The algebraic numbers contain the integers and rational numbers and part of the irrational numbers. The transcendental numbers are the irrational numbers that are not algebraic.

This chapter opened with a comparison of water volumes representing the real numbers versus the natural and rational numbers. The conclusion, based on Cantor's results, implied that the vast volume representing the real numbers is in sharp contrast with the negligible, indeed 0 volume for the natural and rational numbers.

Cantor's results establish an even more astounding conclusion about the algebraic and transcendental numbers. The transcendental numbers, as part of the real numbers, are represented by the entire volume of water, and the algebraic numbers by a volume of 0!

Thus, if we would randomly pick a real number, then with probability 1 it would be transcendental, and it would be a miracle—defined as an event happening with probability 0—that this number would turn out to be algebraic, or more specifically, rational or integer.

So far, we have ignored issues of notation of numbers and simply have used the decimal system. The next chapter shows that notation for numbers, indeed for all mathematical concepts, is not a trivial aspect of mathematics: Notation can propel mathematical development or stifle, even inhibit, it.

3

Notation

Mathematical notation can propel mathematical development or make it nearly impossible. The reason is the structure of the human brain. Due to evolution, it is eminently equipped to look for patterns. Thus, it performs well when the patterns of notation are indicative of underlying concepts, and it may fail badly when this is not the case. In the jargon of computer science, the brain requires user-friendly notation.

This chapter begins with an example case that is like a laboratory study of the brain's capabilities: Two large groups of mathematicians want to create results in a novel area of research. They are given starting information that, mathematically speaking, is equivalent. For one group, the information is encoded in suitable notation that reflects the underlying concepts. The second group is not so lucky; they must use a notation that does not represent the underlying mathematical structure very well. The laboratory study runs over decades. The mathematicians of the first group consistently produce impressive research results, while those of the second group make little progress.

It's hard to believe that something like this could ever happen. Well, it did—not in a laboratory study, of course, but in the real world. The events were exactly as described, with successful outcomes for one group of mathematicians not just over some decades but over a period of 150 years, and mostly failure for the second group during that time. Here is the story.

Newton versus Leibniz: A Controversy

Calculus is concerned with the following two problems.

First, the problem of *differential calculus*: Given a function of a variable, how rapidly does the function change as the variable changes?

Put differently, what is the slope of the function at a given point? Stated in yet another way, what is the slope of the tangent that touches the function at a given point?

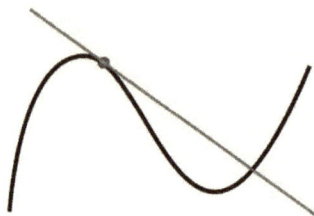

Differential calculus: Tangent straight line touches curved function.[45]

Second, the problem of *integral calculus*: Given a function, what is the size of the area bounded by the function and the horizontal axis?

Here, a certain convention is used. If the function is positive, that is, it lies above the horizontal axis, then the area between the function and the axis is considered to have a positive value. But if the function is negative,

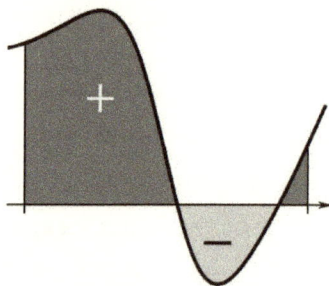

Integral calculus: Area defined by function is total of "+" areas above horizontal axis minus total of "−" areas.[46]

that is, it lies below the horizontal axis, then the area between the function and the axis is considered to have a negative value. The total area is the sum of positive and negative values.

Since ancient times, mathematicians had tried to solve these two problems. They managed to handle special situations, but did not succeed in the general case. In 1666, Isaac Newton (1643–1727) began work on the two problems, and soon found such a method.[47]

PHILOSOPHIÆ

N A T U R A L I S

PRINCIPIA

MATHEMATICA·

Autore J S. NEWTON, Trin. Coll. Cantab. Soc. Mathefeos
Profeffore Lucafiano, & Societatis Regalis Sodali.

IMPRIMATUR·
S. P E P Y S, Reg. Soc. P R Æ S E S.
Julii 5. 1686.

L O N D I N I,

Juffu Societatis Regiæ ac Typis Jofephi Streater. Proftat apud
plures Bibliopolas. Anno MDCLXXXVII.

Left: Isaac Newton, by Godfrey Kneller, 1689.[48]
Right: First edition of Newton's Philosophiae Naturalis Principia Mathematica, 1687.[49]

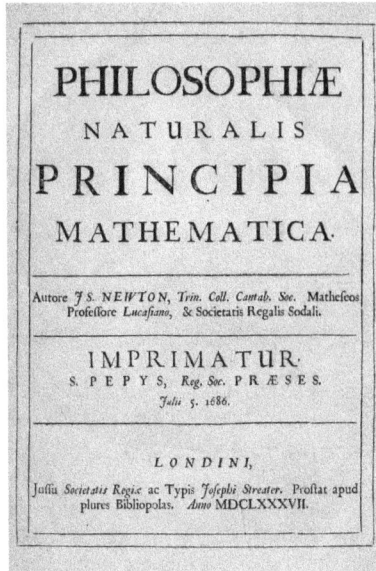

Newton expressed his method using the terms *fluxions* for differential calculus and *fluents* for integral calculus. We shall not go into details here, but mention that these terms were motivated by Newton's concept that changes of quantities represented motion or flow.

Newton used his method extensively in the book *Philosophiae Naturalis Principia Mathematica*, which represented an upheaval in physics. In the three editions of the book appearing in 1687, 1713, and 1726, he described his method[50] for the two calculus problems. Note the delay in publicizing his method: Newton began the work in 1666, but published it for the first time 21 years later!

In a separate development, Gottfried Wilhelm Leibniz (1646–1716) began work on the two problems of calculus in 1674—eight years *after* Newton had begun his effort. For the differential calculus problem, Leibniz defined for a variable x and function $f(x)$ the ratio $\frac{df}{dx}$ where dx is the difference of two x values and df is the difference of the corresponding $f(x)$ values. When dx is chosen infinitesimally small, the ratio constitutes the desired solution. For

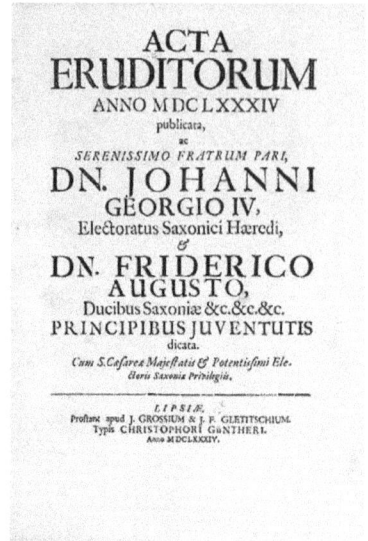

Left: Gottfried Wilhelm von Leibniz, by Christoph Bernhard Francke.[51]
Right: Acta Eruditorum, 1684; contains the first Leibniz publication about calculus.[52]

integral calculus, Leibniz computed the area bounded by the function and the horizontal axis by summing infinitesimal quantities.

Leibniz published his results for the first time in 1684, which means three years *before* publication of the first edition of Newton's Principia. Thus, based on the publication dates, Leibniz had solved the calculus problems first, when in reality Newton had done so earlier.

The story then becomes too complex to be treated here in detail. Initially, Newton and Leibniz acknowledged each other as co-inventors of calculus. But toward the end of the 17th century, misinformation and innuendo by others ignited a bitter feud and ultimately led to Newton and Leibniz accusing each other of plagiarism.

After careful analysis of notes, letters, and publications written by various parties at the time, historians now conclude that Newton first invented calculus, but that Leibniz was not aware of Newton's results and independently invented calculus as well. So, the most fitting conclusion is that Newton and Leibniz are co-inventors.[53]

We draw this conclusion today. But in the 18th century, British mathematicians were convinced that Newton had been grievously wronged by Leibniz. As a result, they rejected Leibniz's approach. In particular, they adhered to Newton's notation, which denotes slope of a function by a dot and integration by a bar.[54]

In contrast, mathematicians on the European continent took Leibniz's side, whose construction uses differences denoted by d and sums denoted by $\int$; the latter symbol represents an elongated "s" of the Latin word "summa." We should emphasize that Leibniz's notation was somewhat different from the modern versions. But the original formulas already reflect the underlying concepts in an intuitive manner.[55]

The superiority of Leibniz's notation quickly became obvious. In a tidal wave of research spanning a 150 year period, continental mathematicians erected a beautiful edifice of ideas and results on Leibniz's foundation. But despite the mounting evidence, British mathematicians continued to adhere to Newton's inferior notation out of loyalty to Newton. As a result, they fell hopelessly behind.

The situation is well expressed by Guillaume de l'Hôpital (1661–1704), who in the first-ever book on calculus[57] writes,

Guillaume de l'Hôpital.[56]

"I must here in justice own (as Mr. Leibnitz himself has done, in Journal des Scavans for August, 1694) that the learned Sir Isaac Newton likewise discovered something like the Calculus Differentialis, as appears by his excellent Principia, published first in the Year 1687 which almost wholly depends on the Use of the said Calculus. But the Method of Mr. Leibnitz's is much more easy and expeditious, on account of the Notation he uses, not to mention the wonderful assistance it affords on many occasions."

The names of the mathematicians building on Leibniz's foundation form an illustrious list. Over the 150 years from 1700 to 1850, it includes, in chronological order, Jacob Bernoulli (1655–1705), Johann Bernoulli (1667–1748), Leonhard Euler (1707–1783), Joseph-Louis Lagrange (1736–1813), Pierre-Simon Laplace (1749–1827), Joseph Fourier (1768–1830), Carl Friedrich Gauss (1777–1855), Augustin-Louis Cauchy (1789–1857), Peter Gustav Lejeune Dirichlet (1805–1859), Karl Weierstrass (1815–1897), and Bernhard Riemann (1826–1866).

It would take too long to even summarize their achievements. But we will review one important problem and its solution. For a moment, let's step back to the time of Newton and Leibniz. A major motivation for the work on differential calculus was the following problem. Given is a function. Where does it attain its maximum and minimum? Differential calculus supplies candidate points for these extrema: They are the points where the slope of the function is 0.[58]

Now consider a more difficult problem: One must find not just maximum and minimum *points* of functions, but must determine *functions themselves* that are best according to some criteria. Here is a problem of that type.

Suppose a straight rail goes from a higher point to a lower point. A piece of metal of some weight is assumed to slide on the rail without friction. Suppose we release the piece at the higher point. Then by the force of gravity the piece will slide to the lower point. We measure the travel time for that movement and wonder: Can we change the shape of the rail so that travel time is reduced? By experimentation, we quickly discover the following. If the rail goes down in a curve with a very steep initial portion, the piece first accelerates quickly and then proceeds with high speed to the lower point. Accordingly, total travel time is reduced.

This fact motivates the following difficult question: What shape must the rail have so that the travel time is as small as possible? The curve achieving the minimum travel time is called the *brachis-*

tochrone curve,[59] from the ancient Greek βράχιστος χρόνος, which means "shortest time." The question thus is: What is the shape of the brachistochrone curve? This question was first posed by Galileo Galilei (1564–1642) in slightly different form. He thought, mistakenly, that the curve should be an arc of a circle.

Left: Galileo Galilei, by Justus Sustermans, 1636.[60]
Right: Johann Bernoulli, by Johann Rudolf Huber, ca. 1740.[61]

Johann Bernoulli was the first to determine the shape of the brachistochrone curve. He then posed the problem as a challenge; Newton, Jacob Bernoulli, Leibniz, l'Hôpital, and Ehrenfried Walther von Tschirnhaus (1651–1708) responded with solutions.[62]

Newton had no difficulty solving the problem. He came home from the Mint, where he had the position of Master, saw that there was a letter from Johann Bernoulli, ate supper, read the letter, solved the problem, and wrote down the solution. All this in one evening. The next day he posted the solution letter to Johann Bernoulli.[63] The story proves that Newton had no difficulties working with his notation. It's just that others could not use it as effectively.

The problem of the shape of the brachistochrone curve and its solution motivated investigation of other problems where best functions had to be found. The research area is now called *calculus of variations.* The mathematicians cited earlier for their contributions to calculus as well as many others created a complex body of theory for this area.[64] The entire effort built upon differential and integral calculus, using Leibniz's notation.

In the second half of the 20th century, the problems of calculus of variations were recast in a novel way, thus greatly simplifying their solution. The key was the *principle of optimality*. It is most easily explained by an example.

Outdoor demonstration: Ring slides down faster on curved brachistochrone rod than on the straight rod.[65]

Suppose we travel from New York to Los Angeles by car via the shortest route. If that route goes through, say, St. Louis, then the principle of optimality says that the portion of the route from New York to St. Louis must be, by itself, the shortest route connecting those two cities.[67]

This age-old and seemingly trivial principle was investigated by Richard E. Bellman (1920–1984), who used it

Richard E. Bellman.[66]

to develop a profound methodology called *dynamic programming*[68] for solution of a variety of optimization problems.

Later, that versatile methodology was combined with the results of differential and integral calculus already known to Newton and Leibniz, resulting in the comparatively easy solution of a large portion of the problems of calculus of variations.[69] The dynamic programming approach is still used today. For example, the US space agency NASA employs it to solve a broad variety of control problems arising in space exploration.[70]

The remaining sections of this chapter look at various cases in the history of mathematics where a new notation or small change of an existing concept led to major developments.

Bürgi and Napier: Multiplication Becomes Addition

In ancient times, complicated notation of numbers made multiplication and division quite difficult.[71] But even with the subsequently developed, much simpler notation of decimal numbers, multiplication and division were tedious, and computation of powers and roots was utterly arduous. Mathematicians found relief in two ways. First, by transforming manual computation using exponents—the subject of this section—and second, by using various mechanical devices, some of which are described in Chapter 7.

The development of the concept of exponents, trivial as it might seem now, took a long time. For example, in the ancient Greek text *Arithmetica* written by Diophantus in the third century,[72] we have the following confusing notation for powers of unknown quantities,[73] where the top row is the modern notation, and the bottom row displays the corresponding expression of the *Arithmetica*.

DIOPHANTI
ALEXANDRINI
ARITHMETICORVM
LIBRI SEX.
ET DE NVMERIS MVLTANGVLIS
LIBER VNVS.
Nunc primùm Graecè & Latinè editi, atque absolutissimis
Commentariis illustrati.
AVCTORE CLAVDIO GASPARE BACHETO
MEZIRIACO SEBVSIANO.V.C.

LVTETIAE PARISIORVM.
Sumptibus SEBASTIANI CRAMOISY, via
Iacobea, sub Ciconiis.
M. DC. XXI.
CVM PRIVILEGIO REGIS.

Arithmetica, 1621 edition, translated into Latin from Greek by Claude Gaspard Bachet de Méziriac.[74]

$$\begin{array}{cccccc} x & x^2 & x^3 & x^4 & x^5 & x^6 \\ \hline \delta & \Delta^Y & K^Y & \Delta^Y\Delta & \Delta K^Y & K^Y K \end{array}$$

At the middle of the 16th century, the mathematician Michael Stifel (1487–1567) writes the following two rows of numbers:[75]

-3	-2	-1	0	1	2	3
$\frac{1}{8}$	$\frac{1}{4}$	$\frac{1}{2}$	1	2	4	8

In the top row, consecutive entries are defined by *adding* 1. Thus, they constitute an *arithmetic progression*. In the bottom row, consecutive entries are produced by *multiplying* by 2. So this is a *geometric progression*.

The numbers are linked as follows. When 2 is raised to the power of a

Michael Stifel.[76]

number of the top row, then the result is displayed by the entry below. Indeed, $2^{-3} = \frac{1}{8}$, $2^{-2} = \frac{1}{4}$, ..., $2^3 = 8$. The "2" in these equations is now called the *base* of the geometric progression. Stifel calls the numbers of the top row *exponents* because they are, well, "exposed." Now comes Stifel's crucial observation:

Suppose we *multiply* two numbers in the bottom row, getting a third number. When we *add* the exponents of those two numbers, we get the exponent of the third number. In modern notation, for any *m* and *n*, we have $2^m \cdot 2^n = 2^{m+n}$.

As a result, if we want to multiply two numbers of the bottom row, we simply add their exponents, go to the position where that sum occurs as exponent, and find the result below that exponent.[77]

Stifel's process simplifies multiplication of numbers to addition of exponents. Similarly, division of numbers becomes subtraction of exponents, taking of powers of a number is reduced to multiplication of its exponent by the specified power, and computation of roots of a number is accomplished by division of its exponent by the specified root.[78]

Stifel established these relationships, but could not use them effectively: After all, the simplified computations could only be carried out if the given numbers occurred in the bottom row. But that row was far from containing all possible numbers.

Toward the end of the 16th century, Jost Bürgi (1552–1632), a master craftsman of clocks and precision machinery as well as an accomplished mathematician,[79] saw a way out of the conundrum faced by Stifel.

Bürgi recognized that Stifel's choice of 2 as base in the definition of the geometric progression was just arbitrary.[82] Actually, any positive number other than 1 would do.

Indeed, if the base was very close to 1, the geometric progression would have very small gaps that could be handled by interpolation. For that reason, Bürgi decided on the base value $B = 1 + \frac{1}{10000} = 1.0001$. He computed the numbers for B^n, $n = 0, 1, 2, \ldots,$ 23027, where the case $n = 0$ corresponds to the number 1.0, and $n = $ 23027 to 9.99999779, which essentially is 10.0. He streamlined the computations so that they likely required just a few months of manual effort.[83]

No additional numbers needed to be computed beyond the cases $n = 0, 1,$ 2, $\ldots,$ 23027 since any positive number of the decimal system is readily

Jost Bürgi.[80]

Mechanized celestial globe, by Jost Bürgi, 1594, Schweizerisches Landesmuseum, Zurich.[81]

scaled by a power of 10 so that it lies between 1.0 and 10.0. Thus, use of the tables only required trivial initial and final scaling by powers of 10.

Alas, Bürgi did not publicize his results in a timely fashion. In fact, he never published the mathematical reasoning behind his method. Instead, in 1620—about 20 years *after* he had completed the research[84]—he published the above described tables[85] and a manual for their use.[86] The manual explains how the tables are used for multiplication, division, and computation of powers and roots. The discussion includes interpolation of table values.

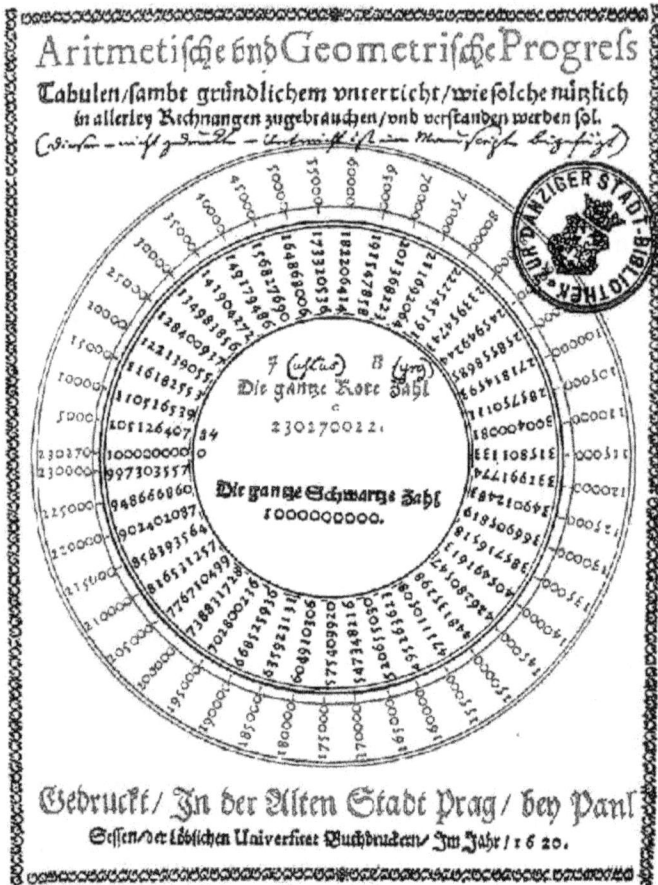

Title page of Jost Bürgi's Table of Logarithms, 1620. The page lists every 500th entry of the table. The numbers are in the inner ring, and the logarithms in the outer ring.[87]

Independently of Bürgi, John Napier (1550–1617)—mathematician, physicist, and astronomer[88]—developed a somewhat different concept of logarithm.[89]

In 1614—six years *before* Bürgi published his manual and tables—Napier presented the results in the book *Mirifici logarithmorum cano-*

nis descriptio. It contained 57 pages explaining the method and 90 pages with logarithm data for the conversion of a number to Napier's logarithm and vice versa.

Left: John Napier.[90]
Right: Napier's Mirifici logarithmorum canonis descriptio, 1614.[91]

The mathematician Henry Briggs (1561–1630) offered[92] to improve the utility of Napier's data by rescaling them to base-10 logarithms. Napier agreed to the transformation. Thus, Briggs carried out the scaling and published the resulting tables. They are sometimes known as *Briggsian logarithms* in his honor.

There have been various discussions whether Bürgi or Napier first developed the idea of logarithm. We shall not weigh in on that controversy.

But it is reasonable to conclude that Bürgi and Napier invented the idea of logarithm independently and approximately at the same time; that Bürgi failed to publish his results in timely fashion and did not work out further details; and that Napier investigated logarithms and their use with extraordinary dedication and care.[93]

Descartes: Geometry Becomes Algebra

In 1637, René Descartes (1596–1650)—outstanding philosopher, mathematician, and scientist[94]—described his research methods for interpreting nature in the book *Discours de la méthode*.[95] In an appendix titled *La Géométrie*, he proposed a revolutionary mathematical concept that allowed treatment of geometric problems by methods of algebra.

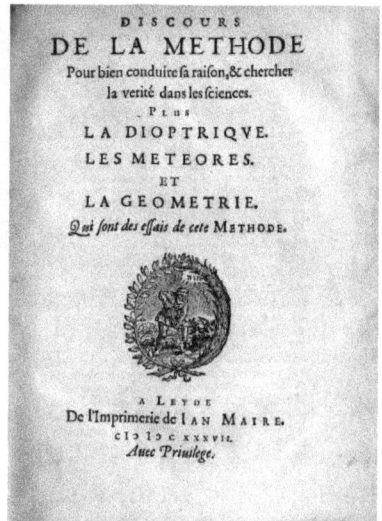

DISCOURS
DE LA METHODE
Pour bien conduire fa raifon,& chercher
la verité dans les fciences.
PLUS
LA DIOPTRIQVE.
LES METEORES.
ET
LA GEOMETRIE.
Qui font des effais de cete METHODE.

A LEYDE
De l'Imprimerie de IAN MAIRE.
CI⊃I⊃C XXXVII.
Auec Priuilege.

Left: René Descartes, after Frans Hals, 1648.[96]
Right: Descartes's Discours de la méthode, 1637.[97]

Up to Descartes's time, geometry had been a favorite area of mathematical investigation since geometric bodies—for example, point, line, parabola, plane, circle, cylinder, cone, pyramid, sphere—naturally embody a number of relationships that do not require explicit mathematical specification.

For example, "sphere" means a 3-dimensional body with a center point from which all points on the surface have the same distance. It also has a surface area and volume.

At the same time, important relationships connect these objects. For example, the angles formed by the lines of a triangle add up to 180 degrees. Due to these facts, complicated results had been proved in ancient times. We shall not go into details of this development,[98] but mention two outstanding results by the ancient Greek mathematician, physicist, engineer, inventor, and astronomer Archimedes (287(?)–212 BCE).

Archimedes, by Domenico Fetti, 1620.[99]

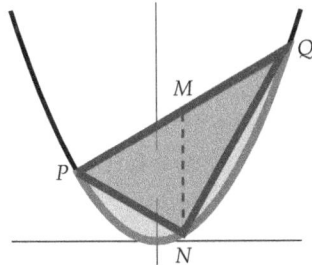

Parabola segment is defined by points P and Q. Triangle is defined by P, Q, and N, where N is vertically below midpoint M of P–Q segment. Area of parabola segment is $\frac{4}{3}$ of area of triangle.[100]

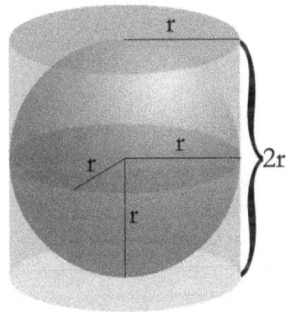

Volume as well as surface area of sphere are equal to $\frac{2}{3}$ that of enclosing cylinder.[101]

The two results plus several others establish Archimedes as the foremost mathematician of antiquity.[102]

The first result concerns a parabola that is cut by a line. The line segment lying within the parabola and a certain point N on the parabola below the line segment define a triangle; see the figure for the definition of N. Archimedes proved that the area within the parabola and below the line is $\frac{4}{3}$ of the area of the triangle.

The second result is the most famous achievement of Archimedes. A sphere is enclosed by a cylinder that is as small as possible. With an amazing sequence of arguments, Archimedes established that the volume of the sphere is $\frac{2}{3}$ of the volume of the cylinder, and that the same ratio applies to the surface areas of the two bodies. Archimedes proved these results with methods that are precursors of the calculus developed 1,900 years later by Newton and Leibniz.

Descartes's *La Géométrie* built a bridge connecting this world of geometry with algebra. To this end, he defined symbols and conventions for constants and variables that are still in use today:

The first letters $a, b, c, \ldots$ of the alphabet denote known quantities, and the final letters $\ldots x, y, z$ represent unknown quantities.[103]

Powers of quantities are specified by superscript, so for any known or unknown quantity q, the expression $q \cdot q \cdot q \cdot \ldots \cdot q$, where q occurs n times, is denoted[104] by q^n.

In *La Géométrie* Descartes also introduced key ideas that others used later to define the *Cartesian coordinate system*.[105] Indeed, he is often credited with inventing that concept, as evidenced by the name "Cartesian."

We describe the system next. Draw a line, declare an arbitrarily selected point to have the value 0, and then demark in equal intervals additional points labeled 1, 2, 3 ... to the right of the 0 and $-1, -2, -3 \ldots$ to the left. Call this line the x-axis. Repeat the construction to create a y-axis. Next, place the x- and y-axes perpendicular to each other in the plane, with the x-axis horizontal and the y-axis

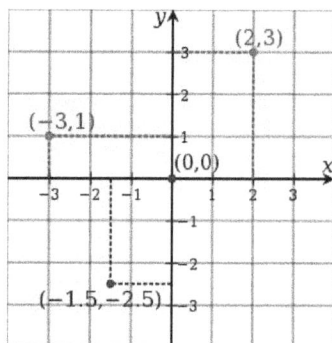

2-dimensional Cartesian coordinate system.[106]

vertical. The axes intersect at their 0 points. The common 0 point is the *origin*.

This configuration is now called the *2-dimensional Cartesian coordinate system*. Any point in the plane is described by its *coordinate values* for x and y as (x, y).

This construction can be extended with another axis, called the z-axis. It is perpendicular to both the x- and y-axes in the obvious way. Now every point of a 3-dimensional world can be defined by a triple (x, y, z) of x-, y-, and z-values in the *3-dimensional Cartesian coordinate system*.

With this machinery, basic concepts of geometry are readily translated into statements of algebra. For example, the points (x, y) of the circle centered at the origin of the plane and with radius r are now defined by the equation $x^2 + y^2 = r^2$. Using such algebraic expressions, results of geometry can be proved much more compactly by algebraic manipulation or by a combination of algebraic manipulation and geometric arguments.

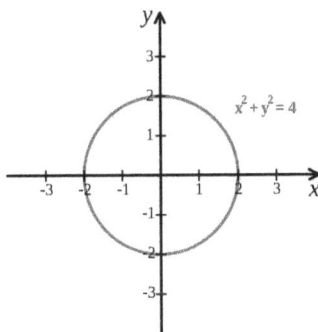

Representation of circle with radius $r = 2$ and centered at the origin.[107]

But Cartesian coordinate systems aren't just a useful device for geometry. Extended to n dimensions by adding axes, they have supported interpretation and manipulation of results in many other branches of mathematics and have become an essential tool for virtually all areas of scientific investigation.

In the next section, we see yet another case where new notation expands the mathematical horizon.

Euler: Formulas Become Functions

Formulas are a core concept of mathematics. When Newton and Leibniz developed calculus, formulas were no longer viewed just as rules of computation, but were analyzed with respect to slope and area.

At that time, Leibniz introduced the term "function" to describe a quantity related to a curve such as slope. The definition of "function" shifted gradually, until in 1755 Euler arrived at a statement that is close to modern usage:

"When certain quantities depend on others in such a way that they undergo a change when the latter change, then the first are called functions of the second. This name has an extremely broad character; it encompasses all the ways in which one quantity can be determined in terms of others."[108]

Euler also introduced the notation "$f(x)$" for functions. With that seemingly simple step, he effectively created a concept of "function" that no longer relied on the specific form of formulas. In the modern definition, this is expressed as follows:

A *function* is a machine that accepts a value of a variable x and outputs the *function value* $f(x)$ for that x value. This concept can be extended to several input variables. For example, the variables may be x and y, and the function is $f(x,y)$.

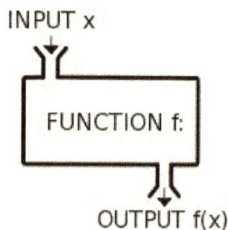

INPUT x

FUNCTION f:

OUTPUT f(x)

Function as a machine with input x and output $f(x)$.[109]

It often is instructive to display the possible input and output values of the machine by a *graph*.

For example, if x and y are the input variables, then the 3-dimensional Cartesian coordinate system with x-, y-, and z-axis can accommodate all input/output cases as points (x,y,z) using $z = f(x,y)$. A listing of the points of the graph is then just another way to specify the function.

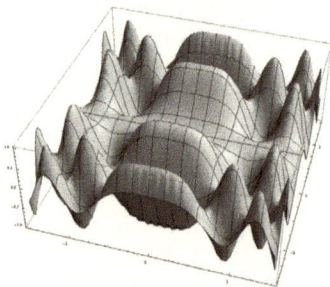

3-dimensional graph of a function $z = f(x,y)$. The x- and y-axes are shown at an angle, and the z-axis is vertical.[110]

Euler's function concept and the $f(x)$ notation shifted the focus from specific rules or formulas to a general view of functions that can be defined, classified, or characterized according to some features.

For example, a function is *continuous* if, informally speaking, it does not jump.[111] Hence, mathematicians began to study continuous functions, defined variations of that property, and looked for conditions that guarantee these versions of continuity to be present.

As a second example, a function $f(x)$ is *differentiable* if for each value of x, the slope of $f(x)$ can be computed. That idea led to conditions that guarantee a function to be differentiable.

The above definition of function also allows for functions for which formulas cannot be specified. On the surface, this may seem like a contradiction: If there is a function, then we must also be able to compute its values, and for that we must have a formula or rule, right? This argument is quite wrong.

In 1936, Alan Turing (1912–1954) proved that there is a function $f(x)$ whose values cannot be calculated by any computer we can imagine. It is called the *halting function*.[112]

The input x of the function is a natural number that represents a computer program.[114] Then $f(x) = 1$ if the program stops after a finite number of steps, and $f(x) = 0$ if the program runs forever. Turing's result implies that there is no formula for the halting function.

Alan Turing, aged 16.[113]

Are there other functions that cannot be evaluated by computer programs? To get some insight into this question, let's take all functions $f(x)$ that for any natural number x have the value $f(x) = 0$ or 1.

Suppose for each such function $f(x)$ we would like to write a computer program that accepts a natural number x as input and then computes $f(x)$. For some functions, this is easy—for example, for the function $f(x)$ that always outputs the value 0. On the other hand, by Turing's result we will be unable to do so for the halting function. Thus we know that we will not be able to write a program for *all* functions. Hence, we modify the goal and write programs for as many of the functions as possible. So imagine that we sit down, pick one of the functions, and try to write a computer program for it. Sometimes we will succeed, sometimes we will fail. Regardless, we then go on to the next function. We stop when we have processed all functions.

For how many functions will we have produced a computer program? For a measure, we take one of the functions at random and ask, "What are the odds that we have written a computer program for this case?" The astonishing answer is that it would be a miracle if we had a program for that function! Mathematically speaking, the probability of having a program for the selected function is 0. An even stronger conclusion applies when we consider functions with slightly different input. That is, the input now is a real number instead of a natural number. The output is 0 or 1 as before. The number of such functions is infinitely larger than that for the functions with natural numbers input and 0 or 1 as output![115]

Evidently, the human mind has created mathematical machinery that is far beyond the need of the practical world of commerce and trade. We saw this first in Chapter 2, where the number of the practically used rational numbers turned out to be an infinitesimal fraction of the number of real numbers.

Euler's work on functions started a virtual flood of mathematical results that looked at functions in various ways.[116] The results justify the claim that the concept of function is one of the most important concepts of mathematics.

Riemann and Lebesgue: Two Ways to Integrate

This section shows how a seemingly trivial shift of viewpoint didn't just expand the power of an already-known mathematical concept, but spawned an entirely new research area. The known concept was integration of functions, and the shift was from vertical slices to horizontal slices of a region.

That shift of thinking led to *measure theory* and the foundation for *probability theory*. The key players in this development were Bernhard Riemann (1826–1866) and Henri Lebesgue (1875–1941).

Left: Georg Friedrich Bernhard Riemann.[117]
Right: Henri Lebesgue.[118]

In 1854, Riemann placed the integration method introduced by Newton and Leibniz on a solid mathematical foundation. Intuitively speaking, Riemann slices the area between the function and x-axis into vertical strips, computes the area for each strip, and then adds up these areas.[119] The process is now known as *Riemann integration*. The method handles many practically important functions, but also fails for a number of instances of interest—for example, when the function jumps so often that any interval of the variable, no matter how small, contains jumps.[120]

Top: Riemann integration.
Bottom: Lebesgue integration.[121]

In 1902, Lebesgue modified Riemann's process in a seemingly trivial way:

He slices the area between the function and the *x*-axis into *horizontal* strips instead of vertical ones.[122]

Lebesgue provided an intuitive description of the process:[123]

"I have to pay a certain sum, which I have collected in my pocket. I take the bills and coins out of my pocket and give them to the creditor in the order I find them until I have reached the total sum. This is the Riemann integral.

"But I can proceed differently. After I have taken all the money out of my pocket I order the bills and coins according to identical values and then I pay the several heaps one after the other to the creditor. This is my integral."

For a demonstration of Lebesgue integration, consider the function $f(x)$ that is equal to 1 if x is rational and equal to 0 if x is irrational. This function cannot be plotted, since the rational and irrational numbers are too closely intertwined. For the same reason, the Riemann integral of $f(x)$ cannot be computed.

In contrast,[124] the Lebesgue integral can be computed, with the conclusion that the area between the function and the *x*-axis is 0, an amazing conclusion![125] When this result is restated in terms of probabilities, we get a proof of the claim of Chapter 2: The probability that a randomly chosen real number turns out to be a rational number is 0.

The main difficulty of Lebesgue integration is measurement of the length of the horizontal strips. Research on that aspect led to an entire theory of measurement. Thus, the seemingly simple change of slicing areas horizontally instead of vertically resulted in a major advance in mathematics.

Summary

We have seen how notation can have a decisive influence on the development of mathematics: Differences of notation hindered or promoted mathematical advances, and seemingly small changes in notation or concepts led to new and profound mathematical results.

The next chapter shows how seemingly reasonable, intuitive thinking about mathematics can be completely wrong. One cause for such failure is the mistaken belief that mathematics is part of nature, that we have a fundamental grasp of nature, and that therefore we have a priori a solid understanding of mathematics that cannot mislead us.

4
Infinity

The human brain has a ready concept for the word "infinity." For example, when the brain hears "an infinite row of trees," it generates a picture of trees neatly arranged in a row extending to the horizon. In the distance, the trees become smaller and smaller until they turn into dots and cannot be discerned anymore. The brain supplements that image with the thought that the trees go on and on beyond the horizon.

The term "infinitesimal" also triggers convincing pictures. For example, upon hearing the definition "A point is a line shrunk to infinitesimal length," the brain envisions a line segment, making it shorter and shorter until it eventually becomes a point.

In the context of poetry or religious texts, the vivid representation of "infinite" and "infinitesimal" causes at most minor problems. In that world, inconsistent information is tolerated, sometimes even purposely introduced, and the brain feels free to create pleasing images.

In mathematics, the brain readily offers the appealing but deceptive belief that "infinity" and "infinitesimal" are actually some sort of numbers.

Sure, these are not regular numbers, but with a bit of care we should be able to link them with the rational or real numbers already on hand. This can indeed be done with suitable restrictions.

But that interpretation can also produce paradoxes and other un-desirable mathematical complications.

In this chapter, we look at instances of such interpretations and their ultimate resolution by entirely different viewpoints. The story begins in the first half of the 17th century, a time of scientific discovery and mathematical innovation, but also of oppression of novel ideas.

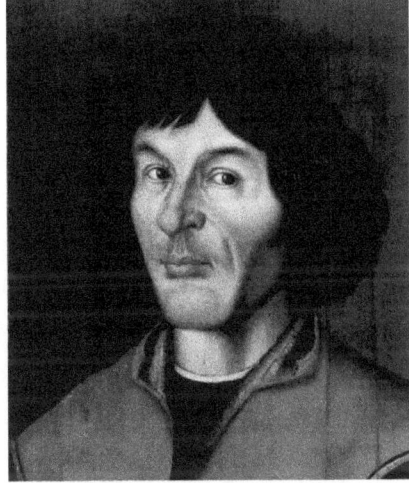

Left: Galileo demonstrates the telescope to Leonardo Donato,[126] the 90th Doge of Venice, and the Venetian Senate, by H. J. Detouche, 1754, detail.[127]
Right: Nicolaus Copernicus.[128]

Galileo, the extraordinary astronomer, physicist, mathematician, and engineer, had learned about the Dutch invention of the tele-scope. Without having seen the device, he constructed one. He was the first person to direct the telescope toward the sky, making as-tonishing discoveries such as the moons of Jupiter, sun spots, and the mountains and craters of the earth's moon. As he scanned the night sky, he became more and more convinced that the ancient model placing the earth at the center was incorrect and should be replaced by the sun-centered model of Nicolaus Copernicus (1473–1543).

At the time, the Jesuits of the Society of Jesus were empowered by the Catholic Church to decide validity of any claim about heaven and earth, be it religious doctrine, philosophical or scientific results, or even theorems of mathematics. The Jesuits roundly rejected Galileo's conclusion, maintaining that the earth was the center of the universe.[129] Threatened with torture,[130] Galileo recanted his views of a universe with the sun at the center and lived under house arrest near Florence until his death in 1642.

In this tumultuous world of controversy and oppression,[131] Bonaventura Cavalieri (1598–1647) and Evangelista Torricelli (1608–1647) laid the foundation for modern calculus.

Cavalieri: Method of Indivisibles

Cavalieri[132] viewed a planar area to be composed of an indefinite number of parallel indivisible line segments, and a body to be composed of an indefinite number of parallel indivisible planar areas.

Cavalieri used these concepts to compute areas of surfaces and volumes of bodies by comparing them with known areas and surfaces.

Left: Bonaventura Cavalieri.[133]
Right: Cavalieri's Geometria indivisibilibus continuorum nova quadam ratione promota, 1635.[134]

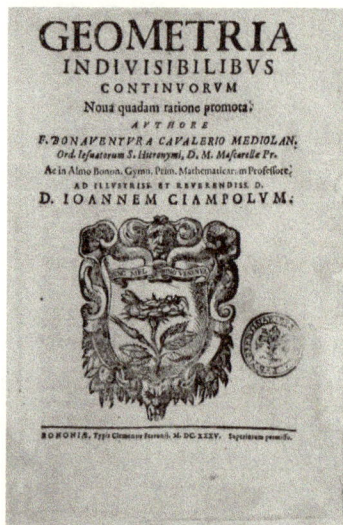

A modern implementation of his method of indivisibles is called *Cavalieri's principle*.[135] It can be stated as follows:

- Suppose two regions of the plane lie between two parallel lines. Further suppose that every line parallel to these lines intersects the two regions in such a way that the line segments falling within the two regions have equal length. Then the two regions have the same area.

- Suppose two bodies in 3-dimensional space lie between two parallel planes. Further suppose that every plane that is parallel to these planes intersects the two bodies such that the intersections have the same area. Then the two bodies have the same volume.

In 1635, Cavalieri published his method in the book *Geometria indivisibilibus continuorum nova quadam ratione promota*[136] (Geometry, developed by a new method through the indivisibles of the continua). At 711 pages, the exposition is very detailed. He communicated extensively about his work with Galileo, who was very much impressed and concluded, "Few, if any, since Archimedes, have delved as far and as deep into the science of geometry."[137]

Galileo's comment referred to the very first methods for integral calculus developed by Archimedes.[138] Expressed in the terminology of Cavalieri, one could say that for each problem Archimedes used a particular collection of indivisibles. Building upon these groundbreaking ideas, Cavalieri created a unifying method using indivisible line segments and areas.

Torricelli: Extension of Indivisibles

Cavalieri's method crucially depends on parallel indivisible lines that have no width and parallel indivisible planes that have no thickness.

In a powerful extension, Torricelli,[139] a student of Cavalieri, considered an indivisible line to be a very thin rectangle, and an indivisible plane to be a very thin slice. This approach allowed the comparison of indivisible lines or indivisible planes that are not parallel. Torricelli's method is best explained by an example.

Left: Evangelista Torricelli, by Lorenzo Lippi, ca. 1647.[140]
Right: Torricelli's Opera geometrica, 1644.[141]

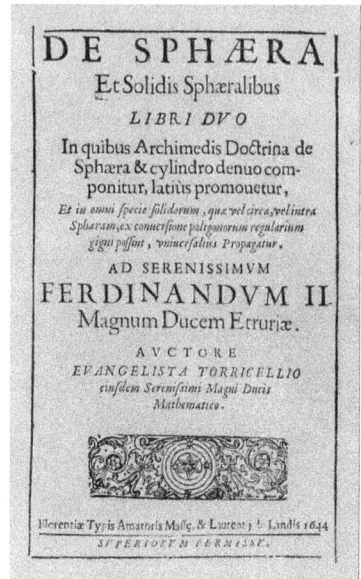

Consider the drawing of an upright rectangle and a slanted parallelogram. They have the same baseline and the same height.

The vertical strip A demarked by the two dashed lines within the rectangle is an indivisible line segment of the rectangle. Corresponding to strip A is the slanted strip B demarked by two dashed lines within the parallelogram.

The two strips share the same portion of the baseline and have the same area. But strip A is shorter than strip B, and correspondingly wider.

Now suppose that a plane region is composed of strips of type A, and that a second plane region has strips of type B in one-to-one correspondence with the strips of type A. Fur-

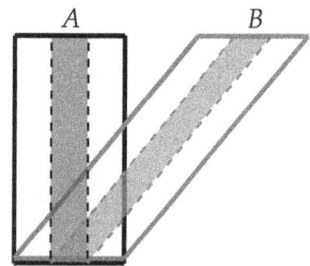

Strip A of rectangle and strip B of parallelogram have same area but different width and length.[142]

ther suppose we know the area of the first region, and that we have a way to compute the ratios between the lengths of the strips of type A versus those of type B.

Then we can compute the area of the second region using the fact of one-to-one correspondence of the two types of strips and the ratios of strip lengths.

Torricelli published his method in the book *Opera Geometrica* (Geometric Works) in 1644. In just 150 pages, it provided a wealth of results that significantly enlarged the foundation for integral calculus.

Among the bodies investigated by Torricelli was a horn-shaped surface now called *Torricelli's trumpet* or *Gabriel's horn*. It is defined via the portion of the hyperbolic curve $f(x) = \frac{1}{x}$ where $x \geq 1$. The trumpet results when that portion is rotated around the x-axis.

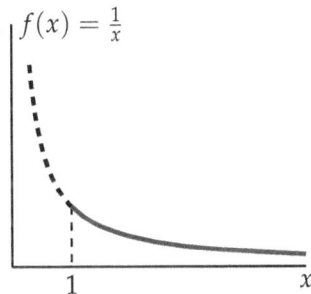

$f(x) = \frac{1}{x}$

Hyperbola. When the solid portion of the curve is rotated around the x-axis, the surface so generated is Torricelli's trumpet.[143]

Torricelli determined that the volume enclosed by the trumpet is finite, but that the surface area is infinite.

It seemed paradoxical that an infinite surface could enclose a finite volume.[145] Since the volume of the trumpet is finite, conceptually it can be filled with a finite amount of paint.

Torricelli's trumpet.[144]

But a finite amount of paint cannot coat the infinite surface of the trumpet, a seemingly contradictory conclusion.

The apparent contradiction is due to an erroneous application of arguments from physics to mathematics: A coat of real-world paint, which has a certain thickness and thus is a 3-dimensional body, is compared with the 2-dimensional surface of the mathematical trumpet.[146]

The work of Cavalieri and Torricelli was strongly supported by Galileo and other mathematicians. But there was a fierce negative reaction by the Jesuits of the Catholic Church.

They continuously monitored research efforts in the sciences and mathematics, looking for consistency with the supposedly eternal truths found in the Holy Scriptures or stated by Aristotle. Indeed, the Jesuits determined that the idea of indivisibles directly contradicted those eternal truths, just as Galileo's claim of a sun-centered universe was heresy.

In a long fight against the use of indivisibles, the Jesuits were ultimately victorious. Coupled with the suppression of Galileo's claim of a sun-centered universe, that victory had a devastating impact on progress in the sciences and mathematics in Italy.[147]

The next step in the understanding of infinity and infinitesimals was taken in the second half of the 17th century in England, far from the oppressive reach of the Jesuits.

Wallis: Infinity and Infinitesimal are Numbers

For the computation of the surface area of a geometrical object, Cavalieri and Torricelli used an indefinite number of indivisible line segments. The term "indefinite" meant that there could be many line segments but that the actual quantity was ignored. The term "indivisible" meant that the width of the line segments could be very small and the exact size needed not be of concern.

The vagueness was well advised:

If the quantity was infinite, then the width couldn't be positive, since such width would make the surface of the given geometrical object infinite. But if the width was 0, the surface area would be 0 as well.

If the quantity was finite, then the width couldn't be very small since otherwise the computed surface area would become very small, too.

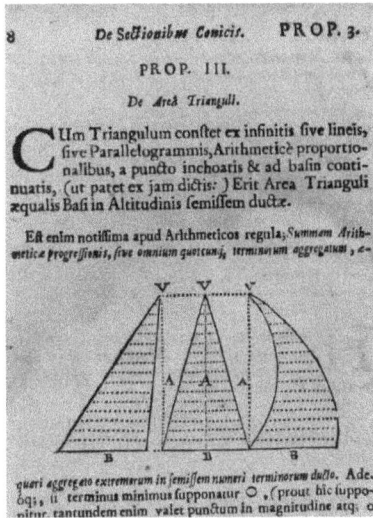

Left: John Wallis, by Godfrey Kneller.[148]
Right: Detail of Proposition 3 of De sectionibus conicis (On Conic Sections) by John Wallis, 1655: three triangular figures with height A and baseline B, each with an infinite number of horizontal strips.[149]

In England, John Wallis (1616–1703) saw this vagueness in the method of indivisibles of Cavalieri and Torricelli and boldly eliminated it by explicit use of infinity and infinitesimal as numbers.[150]

To this end, he defined an infinite number of strips for the computation of surface areas. Each strip had infinitesimal width. He did not try to justify this idea mathematically, but simply argued that his method made intuitive sense and produced correct results.

For computations, he introduced the symbol ∞ for infinity and denoted infinitesimal by $\frac{1}{\infty}$. He used these symbols as if they represented any other number, taking into account that ∞ was very large and $\frac{1}{\infty}$ very small. For example, $\infty \cdot \frac{1}{\infty} = 1$ since he canceled the occurrences of ∞ in the numerator and denominator.[151]

Wallis verified mathematical claims just by checking instances, a generally unacceptable method. But in doing so, he produced impressive results in trigonometry, calculus, geometry, and the analysis of infinite sums.[152] They include a way to compute π with arbitrary precision with the formula $\frac{\pi}{2} = \frac{2}{1} \cdot \frac{2}{3} \cdot \frac{4}{3} \cdot \frac{4}{5} \cdot \frac{6}{5} \cdot \frac{6}{7} \cdot \frac{8}{7} \cdot \frac{8}{9} \cdots$, now known as the *Wallis product*.[153]

Newton and Leibniz: Infinitesimals Become Zero

Cavalieri, Torricelli, and Wallis developed calculus methods for particular classes of problems. The goal of Newton and Leibniz was different: They wanted to solve the general case where a function[154] is given and one desires the slope of the function or the area formed by it and the x-axis.

Newton and Leibniz succeeded admirably. With his calculus method, Newton produced a stunning mathematical foundation of the physical world that stood unchallenged until the time of Einstein. Researchers using Leibniz's approach to calculus developed a huge part of mathematics.[155]

Though Newton's and Leibniz's methods proved to be extraordinarily effective, the arguments supporting them were not fully satisfactory. Perfection would come later, when infinity and infinitesimals were better understood. Let's look at the details of the difficulty.

Consider a function $f(x)$. How rapidly does the function change as x changes? Newton and Leibniz answer the question by considering a small change[156] d of x. The new point is $x + d$, and its function value is $f(x + d)$.

Next, they consider how much the function changes as x becomes $x + d$, that is, $f(x + d) - f(x)$.

Computation of slope.[157]

For example, let $f(x) = x^2$. Then $f(x + d) = (x + d)^2 = x^2 + 2xd + d^2$ and the difference of $f(x + d)$ and $f(x)$ is $f(x + d) - f(x) = x^2 + 2xd + d^2 - x^2 = 2xd + d^2$. At this point, the methods of Newton and Leibniz diverge.

Newton converts the difference $2xd + d^2$ into a *rate of change*, called by him the *fluxion* of $f(x)$, as follows: He first divides the difference

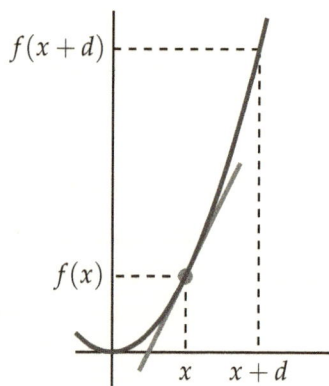

$2xd + d^2$ by d to get the rate of change. This is allowed since d is not 0. The resulting ratio is $\frac{2xd+d^2}{d} = 2x + d$.

Now comes the crucial step. He sets d equal to 0, thus reducing the formula derived for the ratio, $2x + d$, to $2x$. This is the desired fluxion of $f(x)$. To summarize, Newton begins by dividing $f(x + d) - f(x)$ by the nonzero d, but then he evaluates the result by setting $d = 0$! Clearly, the process contains an inherent conflict.

In contrast, Leibniz[158] relies on a hierarchy of infinitesimally small quantities. For the evaluation of the difference $2xd + d^2$, d is considered infinitesimally small, and $d^2 = d \cdot d$ is infinitesimally smaller than d and thus infinite-infinitesimally small.

Based on that conclusion, the term d^2 is declared to be 0, so the difference becomes $f(x + d) - f(x) = 2xd$. Dividing both sides of the equation by d, the slope $\frac{f(x+d)-f(x)}{d} = 2x$ results. Note that the latter approach, while different from Newton's, relies on the same idea that very small quantities can sometimes be set to 0.

In the next section, an entirely different view of calculus eliminates these steps. Why didn't Newton or Leibniz discover that solution?

A reasonable answer is that mathematics was still considered part of nature. Indeed, Newton discussed the dilemma of the supposed division by 0 and appealed to an interpretation of motion for resolution of the difficulty.[159]

Leibniz's work was motivated by the computation of tangents for geometrical figures displayed in the Cartesian coordinate system.[160] He approximated the curves of the figures by piecewise linear segments, then shrunk the segments to infinitesimal size.

From that geometrical interpretation, the idea of several types of infinitely small distances emerged, and it seemed reasonable that infinitely small could be nonzero while infinite-infinitely small was 0.

But contrary to these beliefs, nature simply does not provide a clue how mathematics should deal with infinitesimals or supply a hint how to avoid that concept entirely.

Bolzano, Cauchy, Weierstrass: Continuity

Nature remained a compelling model for calculus and infinitesimals up to the beginning of the 19th century. But then a new view emerged where concepts didn't rely on nature-based arguments, but on unambiguous mathematical terms. Key contributors to this development were Bernard Bolzano (1781–1848), Augustin-Louis Cauchy (1789–1857), and Karl Weierstrass (1815–1897).

They realized that justification of the steps of calculus would be possible only when convergence of sequences as well as smoothness of functions had been precisely defined. We discuss their solution in three steps. First, we cover the definition of convergence of infinite sequences.

Bernard Bolzano.[161]

Step 1: Convergence of Infinite Sequences

Consider the sequence S of fractions $\frac{1}{k}$, $k = 1, 2, 3, \ldots$, that is, $\frac{1}{1}, \frac{1}{2}, \frac{1}{3}, \ldots$. These numbers get ever closer to 0. We express this by saying that the sequence S *converges* to 0, or that 0 is the *limit* of S. The intuitive sense of convergence is captured by the following definition: A sequence S converges to a limit L if its terms eventually get ever closer to that limit.[163]

In the second step, we use the limit concept to characterize so-called continuity of functions.

Augustin-Louis Cauchy. Lithography by Zéphirin Belliard after a painting by Jean Roller.[162]

Step 2: Continuity of Functions

Functions that contain no jumps are called *continuous*. Specifically, a function $f(x)$ is *continuous at a point c* if it doesn't jump at the point $x = c$. It is *continuous* if it is continuous at all points.

There are several equivalent definitions for continuity of functions at a given point c. The simplest definition says: A function $f(x)$ is continuous at a point c if for any sequence of x values, say $x_1, x_2, x_3, \ldots$ that converges to c, the associated sequence $f(x_1)$, $f(x_2), f(x_3), \ldots$ converges to $f(c)$.[164]

In the third and final step, we use the continuity concept to extend functions.

Step 3: Continuous Extension of Functions

Consider the following situation. We have a formula for a function $f(x)$ that defines the function values for all possible values x except for *exactly one* instance, say when $x = c$. At that point, the formula cannot be applied for some reason.

We want to eliminate this gap in the definition of $f(x)$. We also say that we *extend* the definition of $f(x)$ to the point c.

How should we select that extension? A sensible approach is as follows: We try to find a value L for $f(c)$ so that the function is continuous at c.

Karl Weierstrass.[165]

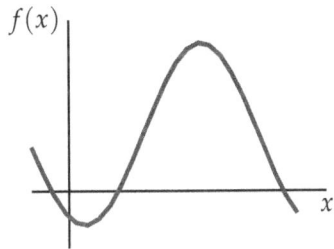

Continuous function.[166]

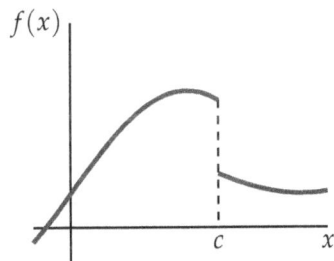

Function with jump at c.[167]

In terms of the continuity definition of Step 2, the selected value L must then observe the following: Whenever a sequence of x values converges to c, then the corresponding function values converge to L.

An abbreviated statement for this latter condition is[168] "the limit of $f(x)$ as x converges to c is L," denoted by $\lim_{x \to c} f(x) = L$.

Thus, the extension process can be written as $f(c) = \lim_{x \to c} f(x)$. We also say that $f(c)$ is the *continuous extension* of the function at c.

The above three steps may now seem simple and almost self-evident. But they were not obvious when Bolzano, Cauchy, and Weierstrass worked to achieve clarity about infinitesimals.

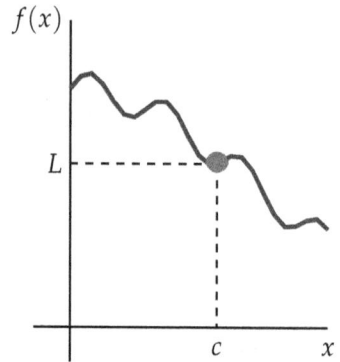

Function $f(x)$ augmented by value L for $x = c$ so that the extended function is continuous at that point.[169]

On the other hand, once these ideas were available, a flood of new insights created a huge area of mathematics now called *Analysis*.[170] It concerns the study of limits, infinite sequences and sums, continuity, differential and integral calculus, calculus of variations, measure theory, and so-called analytic functions, which are defined by certain sums.

We shall not attempt a survey of that area of mathematics. But we do show how the ideas were used to place Newton's and Leibniz's method for the computation of function slope on a solid foundation. We use the example function $f(x) = x^2$ of the preceding section for the demonstration.

Recall that the slope of the function at the point x was estimated via two function values $f(x)$ and $f(x+d)$ as $\frac{f(x+d)-f(x)}{d} = 2x + d$.

Intuitively speaking, we want to eliminate the term d on the right-hand side. Newton did this by setting d directly to 0, while Leibniz eliminated a d^2 term of the intermediate computations. The objection to these steps was, and still is, that such manipulation has no reasonable mathematical justification.

The resolution of the problem relies on the concept of continuous extensions of functions. Define a function $g(d)$ to be the slope estimate for all nonzero values of d; that is, $g(d) = \frac{f(x+d)-f(x)}{d} = 2x + d$. We emphasize that the definition leaves open the value of $g(d)$ at $d = 0$.

Now comes the crucial break with the approaches of Newton and Leibniz. Instead of eliminating d from the formula $2x + d$ by setting d itself or an earlier encountered d^2 to 0, we obtain the as yet undefined $g(0)$ by the extension process.

That is, we look for a value L for the planned extension $g(0) = L$ so that $g(d)$ is continuous at the point $d = 0$. Now for any sequence $d_1, d_2, d_3, \ldots$ that converges to 0, the sequence $g(d_1), g(d_2), g(d_3)$, $\ldots$, which is $2x + d_1, 2x + d_2, 2x + d_3, \ldots$, converges to $2x + 0 = 2x$. Hence, the continuous extension of $g(d)$ at $d = 0$ is $g(0) = L = 2x$.

The above process was done for an arbitrary point x. Hence it is valid for any x value, and we have $2x$ as the slope function of x^2. In modern notation, we have for $f(x) = x^2$ the slope function, or derivative, $\frac{df}{dx} = 2x$.

The same arguments can be used in the general case of slope computation and thus constitute the long-sought remedy for Newton's and Leibniz's setting certain infinitely small quantities to 0.[171]

The remedy also helped identify functions where the slope cannot be computed. In fact, Weierstrass determined a continuous function, now called the *Weierstrass function*, where the slope cannot be computed for any point.[172]

In a similar fashion, limits also replaced indivisibles and eventually led to the Riemann and Lebesgue integration covered in Chapter 3.

Next we turn to Cantor's fundamental work about infinity.

Cantor: Creation and Classification of Infinities

For more than 2,000 years, mathematicians viewed the term "infinity" with suspicion.

Given the common belief that mathematics was part of nature, how could the concept be explained? Clearly, infinity was not a number. So what did it constitute? And how did it manifest itself in mathematical objects?

For some settings, there seemed to be clear relationships between infinities.

For example, it appeared obvious that a line consists of an infinite number of points, a plane of an infinite number of lines, and 3-dimensional space of an infinite number of planes. Therefore, a line had to have far fewer points than a plane, which in turn had to have far fewer points than 3-dimensional space.

As a second example, the rational numbers are ratios of integers, so clearly there had to be many more rational numbers than integers. Indeed, for any two rational numbers r and s, no matter how close, other rational numbers, for example $\frac{r+s}{2}$, lie between them; on the other hand, consecutive integers do not have this property.

As a third example, the algebraic numbers, defined as the roots of polynomials with integer coefficients—see Chapter 2 for details—include all rational as well as many irrational numbers. So clearly, there had to be many more algebraic numbers than rational numbers.

But there were no precise mathematical concepts that on one hand captured the intuitive idea of infinity and on the other hand permitted proof of the above conclusions.

There was a sure way to avoid these troubling thoughts about infinity: Simply demand that the concept of infinity couldn't be used in mathematics. This requirement is now known as *finitism*.[173]

Kronecker believed that finitism should be the underpinning of mathematics. Certainly, by its very definition, finitism was guaranteed to eliminate the conundrums about infinity.

But we know now that it also would have imposed a limitation on mathematical creativity akin to limiting the speed of man-made transportation machinery to that of a pedestrian. That restriction would have converted the automobile to a useless device and limited human flight to balloons controlled by a tether.

Even the ancient mathematicians did not believe that finitism was the correct response to the troubling idea of infinity, and by the 17th century, mathematicians accepted infinity as a valid concept used in various, mostly cautious ways.

Good examples of this development are the results of Archimedes, Cavalieri, and Torricelli. The bold approach by Wallis, who simply declared infinity to be a number, is a notable exception. But despite frequent use of infinity, no mathematically sound concept emerged.

There were philosophical ideas about infinity, for example the concept of *potential infinity* and *actual infinity*.[174]

Potential infinity was exhibited by the natural numbers since they could be enumerated by 1, 2, 3, ... without end. *Actual infinity* concerned the case where the natural numbers were collected in a set, which then could be manipulated. Clearly, that set had infinite size, and this motivated the term "actual infinity."

But these and similarly vague definitions didn't lead to a better understanding of the mathematical use of infinity.

In the 19th century, Cantor bursts into this world of vague infinity concepts with a revolutionary approach. He first solves the following basic problem: How can the concept of size or *cardinality* of a finite set, which simply is the number of elements in the set, be extended to infinite sets?

His solution doesn't rely on direct counting. Instead, it *compares* two infinite sets and results in a relative claim about cardinality. The set N of natural numbers is the fundamental yardstick for this process,[175] since it has the smallest cardinality among all infinite sets under a reasonable assumption.[176]

Cantor denotes the cardinality of the yardstick set N by the Hebrew letter aleph with subscript 0, written $\aleph_0$. Any set T having cardinality $\aleph_0$ is said to be *countable*, due to the fact that the elements of T can be lined up and enumerated just as the natural numbers can be listed as 1, 2, 3, ...

For the cardinality measurement of other infinite sets, the following result is useful. Let R and S be countable sets. Derive a set T from R by, informally speaking, replacing each element of R by a copy of all elements of S. The resulting set T turns out to be countable.[177] Let's call this process *substitution* of S into R to create T.

With substitution, it is easily proved that the sets of integers, of rational numbers, indeed of algebraic numbers, have the cardinality of N and thus are countable.[178]

Another simple proof establishes that the sets of algebraic points contained in a line, plane, 3-dimensional space, indeed any n-dimensional space with finite n, are all countable.[179]

Thus, the line contains just as many algebraic points as any n-dimensional space with finite n. At the time, it was an astonishing upset of conventional wisdom!

Next, Cantor shows that there are more real numbers than algebraic numbers, and thus more real numbers than natural or rational numbers. The key element of the ingenious proof is now called Cantor's *diagonal argument*.[180] It links the real numbers with all possible subsets of the set of natural numbers.[181]

Motivated by this relationship, Cantor denotes the cardinality of the set of real numbers by $2^{\aleph_0}$. How much larger is $2^{\aleph_0}$ relative to $\aleph_0$, the cardinality of the set of natural numbers?

Chapter 2 includes a probabilistic statement that compares the abundance of real numbers, measured by $2^{\aleph_0}$, with the sparsity of the rational numbers, captured by $\aleph_0$: If we choose a real number between 0 and 1 at random, then the probability that it is a rational number is 0. The fact that both the set of rational numbers and the set of algebraic numbers have the same cardinality $\aleph_0$, allows us to expand the claim: The probability that a randomly selected real number is algebraic, is 0 as well.[182]

We include one more result of Cantor. Using a mind-boggling construction of so-called *ordinal numbers*,[183] he defines a set of minimum cardinality that is not countable. He denotes the cardinality of that set by $\aleph_1$.

He now has two uncountable sets that seem closely related: The set of real numbers, with cardinality $2^{\aleph_0}$, and the smallest uncountable set, with cardinality $\aleph_1$. He conjectures that the two cardinalities are the same, so $2^{\aleph_0} = \aleph_1$. If true, the real numbers would constitute a smallest uncountable set.

This conjecture is known as the *continuum hypothesis*. For the rest of his life, Cantor tried to prove this hypothesis, to no avail. Indeed, in Chapter 6 it is shown that no such proof is possible.

Seen against the background of the prior, puny concepts of infinity, Cantor's constructions and claims were an extraordinary expansion of mathematics.

Some mathematicians, in particular those believing in finitism, were aghast: Kronecker went so far as to call Cantor a "scientific charlatan," a "renegade," and a "corrupter of youth."[184]

Looking back, the criticism stemmed from a philosophical misunderstanding of the connection between mathematics and the world: The fact that the concept of infinity is not part of nature does not imply that mathematics should not employ that concept. Cantor simply introduced new axioms about infinity and derived their consequences.

Cantor paid a steep price for publishing his results: Kronecker not only undermined Cantor's relationships with other mathematicians, but also made sure that Cantor could not advance to a coveted professorship in Berlin, Germany. The damage Kronecker inflicted on Cantor is evident from Cantor's letters to Gösta Mittag-Leffler (1846–1927), a friend who

Gösta Mittag-Leffler.[185]

strongly supported him.[186] But eventually Cantor's astounding contributions were fully recognized, as we shall see in Chapter 6.

Summary

The concepts of infinity and infinitesimal stemmed from the realization that the natural numbers go on and on, and that rational numbers can become smaller and smaller.

Over centuries, mathematicians tried to capture and treat these two concepts in various ways. In the process, they produced significant results. But lacking was a firm mathematical foundation that fully justified the conclusions.

Construction of that foundation had to wait till the 19th century, when limits and several types of infinity were invented and became powerful mathematical tools for clarifying the elusive ideas of infinity and infinitesimal.

The struggle for clarity in mathematics is also evident in the attempts to solve seemingly simple but actually very difficult problems that were first posed in antiquity and then resisted solution for more than 2,000 years. The next chapter covers six such problems. Amazingly, all of them were completely resolved in the 19th century.

5
Six Problems of Antiquity

As the ancient Greek mathematicians began to explore the landscape of mathematics created by their imagination, they identified a number of fundamental problems. They proceeded to solve many of them. But some cases, though seemingly solvable, turned out to be difficult and resisted solution.

In this chapter, we look at six of these difficult problems, to be defined in detail shortly: Constructing regular polygons, finding roots of polynomials, trisecting angles, doubling the cube, squaring the circle, and redundancy of Euclid's axiom for parallel lines.

It wasn't just the ancient Greeks who couldn't solve these six problems. Over the next 2,000 years, many mathematicians attempted solutions, got partial results, but could not find complete answers.

Finally, in the 19th century, all six problems were solved in rapid succession. What made this possible? A simple answer would be: Over those 2,000 years, more and more concepts and ideas were developed, and finally there was enough insight to solve the six problems.

But there is a more incisive answer. Up to the 18th century, mathematics was considered part of the world, and mathematical methods agreed with physical concepts. But then mathematics began to stand on its own.

Examples of that development are Euler's invention of the function concept[187] and his declaration that the imaginary numbers, which up to that time were considered to be just figments of the imagination, were numbers as much as the real numbers.[188]

The gap between mathematics and the world grew dramatically in the 19th century. For example, number theory was started by Euclid with the concept of prime numbers and, up to the 19th century, was largely concerned with properties of numbers.

But during the 19th century, profound relationships were established that linked numbers with other, often new, mathematical structures. Cantor's construction of transfinite cardinal and ordinal numbers is an example of that development.[189]

Another shift away from nature occurred in geometry, where the firmly held belief in a world shaped according to Euclid's geometry was shattered by a flood of newly created geometries.

These new and often radical ideas made solution of the six long-standing problems possible.

Definition of the Six Problems

In the real world, one can take a straightedge or ruler, draw a straight line, and then use markings of the straightedge to define distances on the line.

Next, one can take a compass and draw a circle using one of the demarked distances as radius. Alternately, if the compass has angle markings, one can draw circles using those markings.

The ancient Greeks created an abstract version of this process for mathematical investigation. They supposed that the straightedge had *no* markings of distance, and that the compass had *no* device for angle measurement.

Since there were no distance markings on the straightedge, the Greeks also defined an initial line segment of arbitrary length and declared it to have distance equal to 1. Except for that single distance, no other tool for distance measurement was provided. What could be drawn with these elementary tools?

By the way, there is a minor technical point. The ancient Greeks assumed that the compass collapses as soon as it is lifted off the plane. This seems to rule out transfer of distances using the compass. But Euclid showed[190] a

Non-collapsing compass.[191]

construction whereby distances effectively can be transferred by a collapsing compass. Hence, we may assume that the compass does not collapse when lifted off the plane.

From now on it is assumed that all constructions are done with a straightedge without distance markings, a non-collapsing compass without angle markings, and an initial length of 1 defined on a line.

We are ready to state the six problems.

Constructing regular polygons: A regular polygon with $n \geq 3$ edges is obtained from a circle by subdividing the circumference of the circle into n arcs of equal length, then replacing each arc by a straight edge.

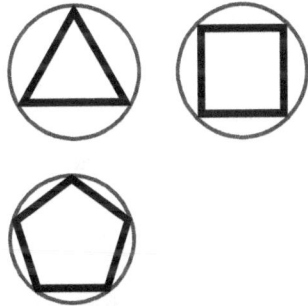

The smallest regular polygons are the equilateral triangle, the square, and the pentagon. The problem demands construction of all regular polygons.

Equilateral triangle, square, and pentagon: the regular polygons with 3, 4, and 5 sides.[192]

Finding roots of polynomials: Chapter 2 discusses the number $\sqrt{2}$ as a solution of the equation $x^2 - 2 = 0$. The left-hand side of the equation is an example of a *quadratic polynomial* with integer coefficients. The general form is $a_2x^2 + a_1x + a_0$, where a_0, a_1, and a_2 are integers. A solution of the equation $a_2x^2 + a_1x + a_0 = 0$ is a *root* of the polynomial $a_2x^2 + a_1x + a_0$. Adding terms, again with integer coefficients, we get the *cubic polynomial* $a_3x^3 + a_2x^2 + a_1x + a_0$, the *quartic polynomial* $a_4x^4 + a_3x^3 + a_2x^2 + a_1x + a_0$, the *quintic polynomial* $a_5x^5 + a_4x^4 + a_3x^3 + a_2x^2 + a_1x + a_0$, and so on. The problem demands the construction of formulas that determine the roots of all such polynomials while using just the four basic arithmetic operations and the taking of the nth root, for any n.

Trisecting angles: Given is an angle defined by two crossing lines. The angle is to be subdivided into three equal angles.

Doubling the cube: Given is a cube. A second cube is to be constructed that has twice the volume of the first one.

Squaring the circle: Given is a circle. A square is to be constructed having the same area as the circle.

Redundancy of Euclid's axiom for parallel lines: Euclid introduced in the book *Elements* five axioms as the foundation of geometry. When the first four postulates are assumed, the fifth axiom is equivalent to the following:[197]

Given a line and a point not on that line, there is at most one line going through the point that does not intersect, and thus is parallel to, the given line.

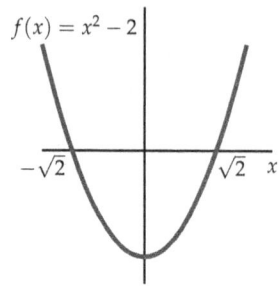

Roots $-\sqrt{2}$ and $\sqrt{2}$ of polynomial $f(x) = x^2 - 2$.[193]

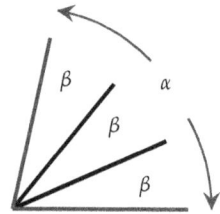

Trisection of angle α into three angles β.[194]

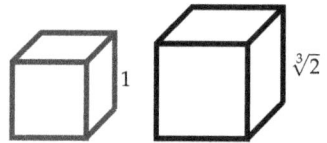

Cube with side length 1 and volume 1, and cube with side length $\sqrt[3]{2}$ and volume 2.[195]

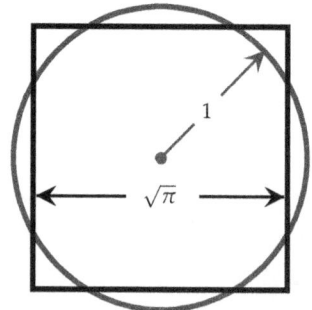

Circle and square with area equal to π.[196]

It seemed utterly obvious that the fifth postulate was implied by the first four. The problem demands that this intuitive idea of the redundancy of the fifth postulate be proved.

Let's see how these six problems were solved.

Given line L and point P: Line M is parallel to L, while line N is not.[198]

Construction of Regular Polygons

The ancient Greeks constructed the triangle, square, and pentagon as well as the 15-sided pentadecagon and the trivial extensions where the number of edges is doubled.[201]

These results provoked the conjecture that maybe all regular polygons could be constructed. But all additional cases resisted solution.

That was the state of knowledge when Carl Friedrich Gauss (1777–1855) considered the problem, at age 19. Gauss arguably became the most eminent mathematician since antiquity.

His subsequent work covered a broad range of areas, including algebra, analysis, astronomy, electrostatics, geometry, geodesy, geophysics, mechanics,

Carl Friedrich Gauss, by Christian Albrecht Jensen, 1840.[199]

Gauss's whimsical signature at age 17.[200]

matrix theory, number theory, optics, and statistics.[202] We get a glimpse of the genius of his work in this chapter.

Gauss had the bold idea to completely disregard the geometric results known at the time and to focus just on five elementary operations on distances that the ancient Greeks had already carried

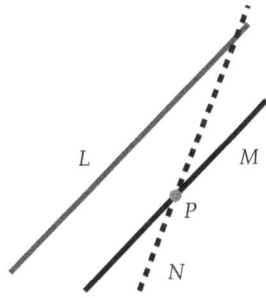

out with geometric constructions: the four basic arithmetic operations of addition, subtraction, multiplication, and division; and the taking of square root.[203]

He also knew that construction of any regular polygon with p corners was equivalent to finding the p roots of the polynomial $x^p - 1$. Indeed, for odd p,[206] the polynomial has one real root, which is $x = 1$, and $p - 1$ complex roots. The p roots are evenly distributed on the unit circle in the complex plane.[207]

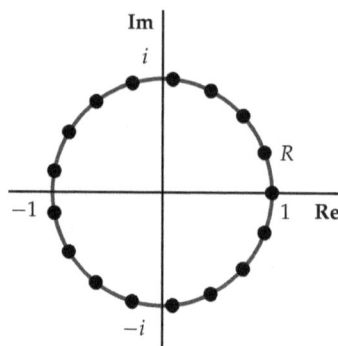

Roots of $x^{17} - 1$ plotted on the unit circle of the complex plane, with real axis Re and imaginary axis Im. Due to the symmetry, derivation of the root $R = \cos(\frac{2\pi}{17}) + \sin(\frac{2\pi}{17})i$ suffices to construct the heptadecagon.[204]

When successive roots are connected by line segments, the regular p-sided polygon results. For the case of $p = 17$, two drawings show the 17 roots and the derivation of the regular 17-sided polygon.

The latter drawing is enlarged so that the difference between circle and polygon becomes apparent.

Gauss describes how he came upon the construction of the regular 17-sided polygon as follows:

Heptadecagon: the regular polygon with 17 sides.[205]

"The history of this discovery [of the construction of the 17-sided polygon] has up to the present nowhere been publicly alluded to by me; I can give it very exactly, however. The day was March 29, 1796, and chance had absolutely nothing to do with it. Before this, indeed during the winter of 1796 (my first semester in Göttingen), I had already discovered everything related to the separation of the roots of the equation[208] $\frac{x^p - 1}{x - 1}$ into two groups[209].... After intensive consideration of the relation of all the roots to one another on arithmetical grounds, I succeeded during a holiday in Braunschweig, on the morning of the day alluded to (before I had got out of bed), in viewing this relation in the clearest way, so that I could immediately make special application to the 17-side[d] [polygon] and to the numerical verification."[210]

It was the first new construction of a regular polygon since antiquity. Gauss's radical break with history made it possible: He used the complex roots of a polynomial and replaced geometric steps requiring straightedge and compass with the basic operations of arithmetic and the taking of square roots.

Five years later, Gauss constructed an entire collection of regular polygons. His result uses the *Fermat primes* first investigated by the mathematician and lawyer Pierre de Fermat (1607–1665).

They are of the form $2^{(2^n)} + 1$, for $n \geq 0$. Only five Fermat primes are known: 3, 5, 17, 257, and 65537, corresponding to the cases $n = 0, 1, 2, 3,$ and 4. Indeed, so far there is no proof establishing either existence or nonexistence of additional Fermat primes.[211]

Pierre de Fermat.[212]

Here is Gauss's result: A regular polygon is constructible by straightedge and compass if the number of edges is a product of distinct Fermat primes and powers of 2.

He conjectured that the condition was also necessary. Pierre Wantzel (1814–1848) proved that conjecture 36 years later.[213] Thus, the polygons specified via Fermat primes and powers of 2 are precisely the constructible regular polygons.[214] This result is now known as the *Gauss-Wantzel theorem*.

Gauss valued the construction of the heptadecagon so much that he requested it to be carved on his tombstone. The stonemason declared that this was impossible[215], since it essentially would look like a circle. A compromise was found for Gauss's memorial in Braunschweig: It shows a star with 17 points.

Star with 17 points at Gauss Memorial, Braunschweig.[216]

Finding Roots of Polynomials

Since antiquity, mathematicians tried to find formulas for the roots of various polynomials. This turned out to be easy for the quadratic polynomials, as we are taught in high school.

The roots for cubic polynomials were much harder to determine, but were eventually found in steps that started with special cases and terminated with the most general solution in the 16th century.[219] The roots of the quartic polynomials were also found in the 16th century by reduction to the cubic case.[220] But the quintic case, where the polynomial generally is of the form $a_5x^5 + a_4x^4 + a_3x^3 + a_2x^2 + a_1x + a_0$, could not be solved despite considerable effort. Then, in the 19th century several mathematicians published results that the quintic case as well as all higher cases were generally unsolvable. Here is a sketch of the developments.[221]

Paolo Ruffini.[217]

In 1799, Paolo Ruffini (1765–1822) established the result that the quintic and all higher cases in general could not be solved. The proof was incomplete.

Niels Henrik Abel.[218]

In 1824, Niels Henrik Abel (1802–1829) published that same result. The claim contained a flaw that, in hindsight, is not considered major.

In 1845, Wantzel acknowledged the prior work of Ruffini and Abel while publishing another proof. Today, Abel and Ruffini are credited with the result, now known as the *Abel-Ruffini theorem*.

Unaware of all these developments, Évariste Galois (1811–1832) proved

Évariste Galois, age 15.[222]

the result in 1829, at age 18. It was published posthumously in 1843.

Hidden behind this terse summary is a story of misfortune of Abel and Galois worthy of a Greek tragedy. Mismanagement of the submitted papers and bungling of referees prevented timely publication of the papers and proper recognition of Abel and Galois during their short lives. Abel died at age 26 in poverty, his results largely unrecognized; two days later, a letter arrived offering him a professorship in Berlin. Galois died at age 20 from wounds suffered in a duel.[223]

Abel and Galois investigated algebraic manipulations with a new concept now called *group*. A group consists of a set of elements and two operations that are inverse to each other. For example, the set of integers and the operations of addition and subtraction define a group.[224].

A *field* is a more complex structure. It involves a set of elements and four operations. For example, the rational numbers with addition, subtraction, multiplication, and division with nonzeros form a field.

When the set of elements is finite, the group or field is called finite.[225]

Galois created a theory, now called *Galois theory*, that links certain groups with fields. To honor Galois, the finite fields are called *Galois fields*.[226] For each of them, the number of elements is equal to a power of a prime number.

Taken together, the ideas of groups and Galois theory constitute major steps of the 19th century separating mathematics from the real world.

Trisection of Angles

Trisection of angles demands that an arbitrary angle is subdivided into three equal angles using straightedge and compass. Wantzel compared the operation of trisection with the construction steps of straightedge and compass, using polynomials to represent the steps, and thus showed that the construction steps couldn't possibly trisect all possible angles.[227]

His proof that regular polygon construction is possible only when the number of edges is a product of distinct Fermat primes and powers of 2—see the *Gauss-Wantzel theorem* described above—also implies that trisection generally is not possible.[228] The trisection problem can also be solved using Galois theory.[229]

Once more we have seen how new abstract concepts made solution of a long-standing problem possible, demonstrating again how mathematics of the 19th century moved away from concepts of the world. The same conclusion applies to the solution of the next ancient problem.

Doubling the Cube

For the discussion, it suffices that the given cube has side length $x = 1$, and thus has volume $x^3 = 1$. Doubling that cube requires constructing a cube with side length y such that the volume is $y^3 = 2$. Evidently, the side length must be $y = \sqrt[3]{2}$.

Wantzel proved that this was impossible using polynomials, just as he had argued in the angle trisection problem.[230] The result can also be obtained via Galois theory.[231]

The fifth problem, which requires the squaring of the circle, turned out to be much more difficult.

Squaring the Circle

Commerce and trade of ancient times often involved computation of areas of geometric figures. The simplest case was the square: To obtain the area, one only needed to multiply the side length with itself.

The ancient Greeks posed the following general problem: Given a geometric figure in the plane, derive a square with the same area. The problem was called *squaring* the geometric figure.

A number of cases were easily solved. For brevity we will not cover the geometric steps, but simply show, equivalently, that the area could be computed by the four basic arithmetic operations and the taking of square root, just as Gauss did when he constructed the heptadecagon. Once the area is known, taking the square root gives the side length of the equivalent square.

Case of the *triangle*: A simple formula, known since ancient times, says that the area is one half of the length of any side of the triangle times the height of the triangle for that side. The height can be constructed, and thus the area be computed.

Case of any *polygon*, which is a geometric figure in the plane whose boundary consists of straightline segments: Divide the polygon into triangles, and sum up the areas of the triangles.[233]

Case of figures with a curved boundary: In antiquity, the squaring of any

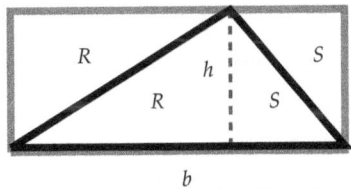

Triangle with baseline b and height h consists of areas R and S. Rectangle has two Rs and two Ss, and total area is $b \cdot h$. Hence triangle area is $\frac{b \cdot h}{2}$.[232]

such figure was considered very difficult.

Hippocrates of Chios (470(?)–410(?) BCE) created the first result for such a geometric figure, now called the *Lune of Hippocrates*. He proved that the lune has the same area as the triangle, and thus can be squared.

The proof mainly consists of a single application of Pythagoras's theorem for right triangles.[234]

Archimedes solved a much more difficult problem: He squared any parabola segment. Indeed, he proved that the area of a given parabola segment is $\frac{4}{3}$ of a certain inscribed triangle. Thus, one only needs to construct the triangle and compute $\frac{4}{3}$ of its area.[238]

Archimedes's method was a precursor of integral calculus. As that calculus was gradually developed,[239] the ancient methods for squaring geometric figures became increasingly unimportant, with one exception: the squaring of the circle.

That problem continued to fascinate mathematicians, for the simple reason that squaring seemed obviously possible if one could just find the correct approach. The 19th century brought clarity to this famous problem. For the discussion, we need the concept of real, algebraic, and transcendental numbers intro-

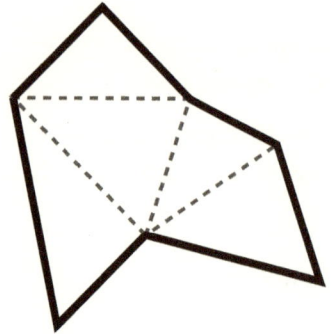

Partition of polygon into triangles.[235]

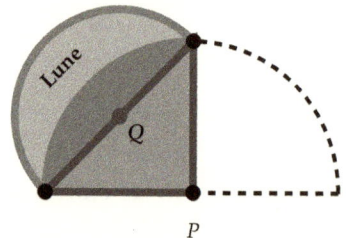

Lune of Hippocrates is defined by quarter circle centered at *P* and semicircle centered at *Q*. It has the same area as the triangle.[236]

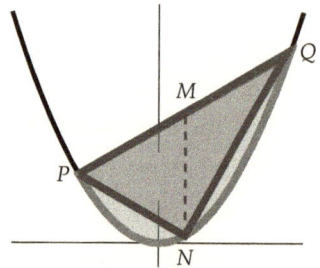

Archimedes: Parabola segment is defined by points *P* and *Q*. Triangle is defined by *P*, *Q*, and *N*, where *N* is vertically below midpoint *M* of *P*–*Q* segment.[237]

duced in Chapter 2. Recall that the algebraic numbers are the real numbers that are roots of polynomials with integer coefficients. The transcendental numbers are the real numbers that are not algebraic.

Let's consider squaring of the circle with radius $r = 1$. Its area is $r^2\pi = 1^2\pi = \pi$, and the equivalent square has side length $\sqrt{\pi}$. Squaring the circle then amounts to constructing $\sqrt{\pi}$, or equivalently due to multiplication, π.

Wantzel showed that the four basic arithmetic operations and the taking of square root could only produce numbers that were roots of polynomials with integer coefficients. Indeed, the exponents of those polynomials were even.[240] Thus, all numbers produced by that process are algebraic.

This implies that a squaring method producing π with these operations could only exist if π was algebraic. The hope that π had this property was dashed in 1882, when Lindemann showed π to be transcendental.[241]

Thus, squaring of the circle was finally proved to be impossible. That result was based on the vastly expanded knowledge about numbers and functions developed in the 19th century.

All problems discussed so far effectively call for geometric constructions or algebraic formulas. The sixth and last problem is different. It demands insight into the role of Euclid's axioms for plane geometry.

Redundancy of Euclid's Parallel Axiom

In his book *Elements*, Euclid introduced five postulates for plane geometry.[242] When the first four postulates are assumed, the fifth is equivalent to the following axiom[243] due to John Playfair (1748–1819):

In a plane, given a line and a point not on it, at most one line parallel to the given line can be drawn through the point.

There are several other axioms that are equivalent to Euclid's fifth.[244] The discussion below also relies on the following:

The sum of the angles in every triangle is 180 degrees.

Mathematicians desire systems of axioms to be minimal; that is, no postulate should be implied by the others.[246] In the case of Euclid's postulates, it seemed utterly obvious that the parallel postulate was implied by the first four. But for 2,000 years, no-

John Playfair, by Henry Raeburn.[245]

body could prove this supposedly self-evident fact.[247] Indeed, when one looks at the drawing demonstrating Euclid's parallel and nonparallel lines, it seems that one can never come up with another system of lines where parallel lines behave differently. The mesmerizing effect of this image was overcome in the 19th century, when a new geometry was proposed that did not use Euclid's fifth axiom.

The history of that new geometry is too complicated to be covered here with reasonable precision. However, the history was compiled[248] with great care at the end of the 19th century when letters and notes of the various contributors were still available. The following summary is based on that material.

At the turn of the 19th century, Gauss was already aware that Euclid's fifth axiom could not be proved and that a different geometry was possible where the sum of angles of any triangle was less than the 180 degrees guaranteed by Euclid's fifth axiom. He was reluc-

Ferdinand Karl Schweikart.[249]

tant to publish this material, fearing a backlash against the new idea. Indeed, he kept quiet about his insight except in letters.

In 1818, he became aware of work by Ferdinand Karl Schweikart (1780–1857), who had concluded that another geometry was possible that he termed *astral geometry*.[250]

In comments about Schweikart's material, Gauss indicated that he was aware of the results.[251] He wrote, "Es ist fast alles mir aus der Seele geschrieben," which was a polite way of saying, "Almost all of it is familiar to me."

In 1824, Franz Adolph Taurinus (1794–1874), a nephew of Schweikart, sent some results about geometry to Gauss. In a long and kind response, Gauss wrote that the results were enjoyable to read, but also pointed out that they were quite incomplete. He added that for more than 30 years he had worked on non-Euclidean geometry, and concluded by requesting—indeed demanding—that Taurinus should view the letter as a private message that he could not publish or reference.[252]

In 1829, Nikolai Lobachevsky (1792–1856) independently developed the new geometry, which eventually came to be called *hyperbolic*. He replaced Euclid's fifth axiom by the following:

For any given line and any given point not on that line: The plane containing both the line and the point also contains at least two distinct lines through the given point that do not intersect the given line.

Nikolai Lobachevsky, by Lev Kryukov, ca. 1843.[253]

There are several ways to display the hyperbolic plane.

Henri Poincaré (1854–1912) proposed a beautiful method, as follows. The hyperbolic plane is represented by a circular disk D with radius 1 in the Euclidean plane.

The hyperbolic points are the points of D that do not lie on the boundary of D, such as point P.

The hyperbolic lines are represented by arcs of circles that lie within D and make a right angle with the boundary of D, such as lines B and C, or by lines that are diameters of D, such as line A.

Lines B and C go through the point P and do not intersect with the line A. According to the definition of "parallel," the lines B and C are thus parallel to line A.[256]

The display preserves angles, and by inspection we can tell that example triangle T has angles summing to less than 180 degrees.

This is entirely different from Euclidean geometry, which can be displayed with straight lines and with exactly one parallel line for any given point outside a given line.

Lobachevsky published his results in Russian; mathematicians outside Russia became aware of his work only years later.

In 1832, János Bolyai (1802–1860) also developed the hyperbolic geometry, again independently.

When Gauss became aware of Bolyai's result, he declared that he had thought of this result decades ago,[258] but also

Jules Henri Poincaré.[254]

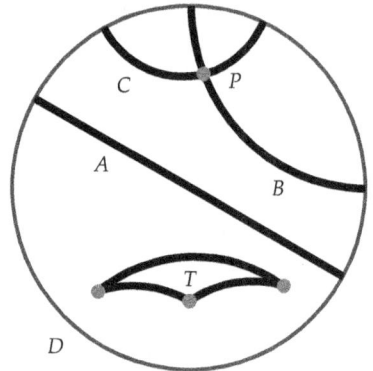

Poincaré's disk D: A, B, and C are lines. T is a triangle.[255]

János Bolyai.[257]

wrote to a friend,[259] "I regard this young geometer Bolyai as a genius of the first order."[260]

Now and then the question is raised how much Gauss had established about the hyperbolic geometry prior to seeing Bolyai's results. Gauss's letters cited above are clear evidence that he fully understood the new geometry. Thus, it is appropriate that Gauss, Lobachevsky, and Bolyai are generally considered to be co-inventors of the hyperbolic geometry.

For a time, it was not clear whether the definition of the hyperbolic geometry contained some inconsistency; that is, whether there was some contradiction inherent in the axioms.

Gauss was troubled by this question, and this likely was one of the major reasons why he did not publish his results.[261]

In 1868, Eugenio Beltrami (1835–1900) resolved this important question[263] by proving that either both Euclidean and hyperbolic geometries were consistent or both were inconsistent.

This result put to rest the lingering suspicion that, somehow, the hyperbolic geometry was inherently flawed. If so, that fate was shared by Euclidean geometry.

The proof that hyperbolic geometry was just as valid as Euclidean geometry eliminated a mental barrier about the nature of geometry.

Eugenio Beltrami.[262]

No longer was there a single geometry that governed all: A change of just one axiom of the Euclidean geometry had resulted in a new, equally acceptable framework.

That insight led to a flood of new geometries, beginning with Riemann in 1853. He expanded the concept of geometry to what is now called *Riemannian geometry*. The hyperbolic geometry is a special case of Riemannian geometry, as is the *elliptic geometry*, where all lines intersect.[264] The elliptic geometry can be visualized on the surface of a sphere.

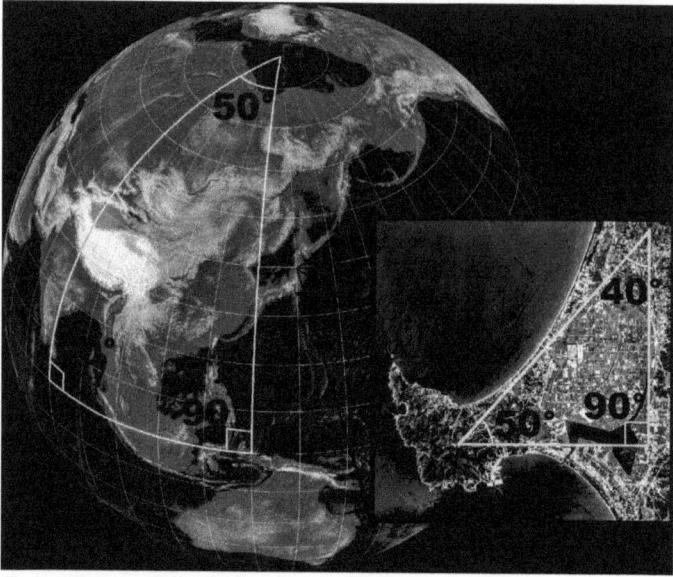

The earth as a spherical model of elliptic geometry.[265]

The great circles of the sphere, which are the circles dividing the sphere into two equal hemispheres,[266] represent the lines.

Any two *antipodal* points, which are points on the sphere opposite to each other, represent together just one point of the geometry.[267]

In that geometry, the sum of angles of a triangle is always greater than 180 degrees. For example, in the large triangle drawn on the earth as an example sphere, the angles sum to 230 degrees. As the triangle becomes smaller and smaller relative to the size of the sphere, the sum of the angles gets ever closer to 180 degrees, as is demonstrated by the triangle in the inset of the picture.

Summary

Six problems of antiquity, open for more than 2,000 years, were all solved in the 19th century. The main reason for that success can be traced to a fundamental change of viewpoint, initiated by Euler in the 18th century: Mathematics is different from nature, does not need nature, and should not be confused with nature.

That thought developed gradually. It met considerable resistance at times, but ultimately prevailed. The results were an array of new mathematical concepts such as groups and fields, and new geometries.

Coupled with new levels of abstraction—for example, the transfinite cardinal and ordinal numbers—this development was nothing short of revolutionary.

———————

Up to this point, we have carried out arguments about mathematical ideas and claims as if there were a natural way to prove results. The next chapter shows that this assumption about mathematical arguments, made since ancient times, has no a priori justification.

Starting in the 19th century, mathematicians made an extensive effort to remedy this misconception. As we shall see, there were major successes, but also humbling failures.

6

Proof

Every day, we face a tsunami of information pouring from newspapers, radio, TV, and the Internet. Much of it is just somebody's guess or, worse yet, artful lying.

In response, we reject information unless there is some proof of veracity, such as: A trusted person confirms the information; the data include the results of scientific tests that support the claim; we have knowledge about the subject matter, and the statement is consistent with that information; pictures validate the message; or a trusted website[268] confirms correctness.

Of course, these checks don't really prove validity. They just make it more likely that the information is correct.

Scientists have erected a higher hurdle for proofs of validity; they demand confirmation of claims by repeated trials.

Coupled with Occam's razor[269]—it prefers simpler explanations if there is a choice—scientists have an outstanding track record of weeding out incorrect results and confirming valid ones.

As an aside, the method does have a drawback: Nature may produce very rare events that, even if recorded by chance, will be rejected since they do not recur in subsequent experiments. Thus, science may be creating a picture of a more evenly behaving nature than is actually the case.

Due to experiment-based evaluation, scientists readily modify or even abandon a model or theory when contrary experimental results show up.

For example, in the world of Newton, space extends evenly in all directions, and time flows evenly forever.

For 250 years, that view was considered correct. But in the early 20th century, Albert Einstein (1879–1955) replaced it with the model of *spacetime*[272] where gravity bends space and time depends on the relative speed of the observer.

Albert Einstein.[270]

The model is a generalization of the Riemannian geometry of Chapter 5. It has been confirmed by numerous experiments, some of which are ongoing today.

When a scientific theory has been confirmed by numerous experiments conducted over long periods of time, the theory may be *postulated* to be an infallible *law of nature*.

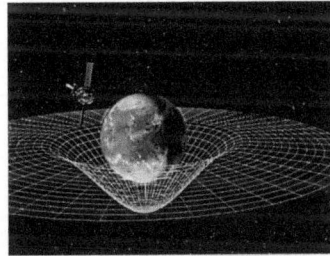

Illustration of a NASA probe orbiting the earth to measure space-time. Note the bending of space by gravity.[271]

For example, when water is heated sufficiently, it starts to boil and becomes steam. This conversion of liquid to steam has been observed for thousands of years, so we have postulated that it is a law of nature.

In mathematics, the bar for proof is even higher: Results cannot be verified just by a number of experiments, but must be shown to be *permanently valid*. Put differently, mathematical results must be correct independently of time and the state of the world. This goal is laudable, but can it be achieved? Two problems stand in the way.

First, any mathematical result in one way or other relies on some initial assumptions and some deductive methods. If the selected assumptions or deduction methods are in dispute, then some mathematicians accept the result while others reject it. So what does "valid" mean in this context?

Second, the precision demanded by mathematicians for assumptions and deductions may increase over time; indeed, it has done so for centuries. Accordingly, a generally accepted result may later—maybe after centuries—be viewed with suspicion. At that time, the result is proved again according to the higher standard, or, if that cannot be achieved, declared to be unproven. However, even if the result is proved again, how can we be sure that the new proof won't be found to be defective at some time in the future?

Early on, mathematicians were aware of these two problems and tried to avoid them with varying degrees of success. By the mid-19th century, enough insight had been gained that mathematicians dared a frontal attack on the problems of precision and validity of results.

That effort lasted roughly 100 years. It produced satisfying insight on many fronts, but also the disturbing conclusion that some results were forever unprovable.

Progress was not always accompanied by pleasant discussions. Indeed, at times there were vehement disagreements that separated mathematicians into camps, with each group claiming perfect insight and correctness. This chapter's main focus is on those 100 years of mathematical developments. They are yet another demonstration of human ingenuity and persistence.

To set the stage, we take a brief look at the proof techniques of antiquity.

Ancient Babylonians: Proof via Drawings

The ancient Babylonians faced a major hurdle when they tried to prove a result. They knew integers and rational numbers, but did not have the modern concept of variable or formula. Nevertheless, they proved interesting results by focusing on geometric problems.

Clay tablet YBC 7289 of 1800–1600 BCE, already discussed in Chapter 2, depicts one such result. The tablet declares a precise approximation to be the length of the diagonal in a square with side length 1. In equivalent decimal notation, it is 1.41421297.

Clay tablet YBC 7289, ca. 1800–1600 BCE.[273]

Another example is clay tablet Plimpton 322, written around 1800 BCE. It lists integer triples a, b, and c that are solutions to the equation $a^2 + b^2 = c^2$.

For a right triangle whose sides have the lengths a, b, and c, with c largest, Pythagoras's theorem[275] establishes this equation. For this reason, triples satisfying the equation are called *Pythagorean*.

Clay tablet Plimpton 322, ca. 1800 BCE. Lists Pythagorean triples.[274]

Rudman argues convincingly[276] that the Babylonians most likely used geometric figures to derive the triples. Based on that analysis, he concludes that the Babylonians knew of Pythagoras's theorem about right triangles at least 1,200 years before Pythagoras.

During the period 350–50 BCE, the Babylonians carried out precise astronomical computations with geometrical methods, a fact recently discovered by Ossendrijver[277] while analyzing several Babylonian clay tablets. Until that discovery, it was assumed that the astronomers of Babylon solely used arithmetical methods and not geometrical ones.

The key result, shown on the next page in modern notation, is as follows. The velocity of the planet Jupiter is plotted over a time interval of 120 days. Time 0 is the point where Jupiter rises over the horizon, and time 120 is the point of *first station*,[278] where Jupiter seems to stand still before beginning the *retrograde motion*.[279] The total area under the velocity curve then represents the displacement of the planet, measured in degrees, during that time interval.

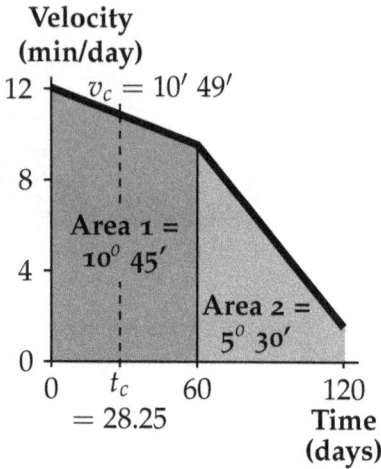

Left: Velocity graph for Jupiter from the time the planet rises on the horizon to the time of first station. The velocity is expressed in minutes per day, where 60 minutes are equal to 1 degree.[280]
Right: Clay tablet BM 34757 of 350–50 BCE contains trapezoid computations of area 1 of the figure.[281]

The figure divides the area under the velocity curve by a vertical line at time 60 into two trapezoids.

The computations for the left trapezoid are recorded on clay tablet BM 34757. During the time period of that trapezoid, Jupiter moves 10^0 45'. During the time of the right trapezoid, Jupiter moves 5^0 30'.

The symbol t_c on the time scale denotes the time when the planet has covered half of the left trapezoid area. The velocity at that time is v_c. The Babylonians computed quite precise values for t_c and v_c, given that the exact solution involves nonterminating fractions of their hexadecimal system of numbers; the value 28.25 of t_c is a rounded decimal. The methods predate related techniques of medieval European scholars by at least 14 centuries.[282]

We move on to the groundbreaking work of the ancient Greeks.

Ancient Greeks: Logic, Axioms, Proofs

Aristotle (384–322 BCE) formalized the logic of deduction in several treatises. Collectively, they are called the *Organon*, which in Greek means "instrument" or "tool." For 2,000 years, this work was considered to be a complete treatment of logic that would never require substantive change or expansion.[283]

In his book *Elements*, Euclid assembled axioms of geometry and, starting with those basic assumptions, proved essentially all results of geometry known at that time.[284]

Archimedes created an array of amazing results that motivated the work of mathematicians for the next 2,000 years.[286] Examples are included in Chapters 3 and 5.

Viewed together, the achievements of just these three mathematicians constitute a profound foundation of mathematics: The results provide a framework for reasoning in proofs, show

Aristotle. Marble, Roman copy after a Greek bronze original by Lysippos from 330 BCE; the alabaster mantle is a modern addition.[285]

how axioms should be formulated and used, and demonstrate that intricate proof techniques can produce astonishing results.

Of course, there were a number of other accomplished mathematicians of antiquity who contributed to fundamental insights. Due to space limitations, we cannot possibly discuss them or their results even in a cursory manner.[287]

Up to the 17th century, the logic of Aristotle stood virtually unchanged. But then Leibniz started a research effort that created many elements of modern logic. Unfortunately, he never published

his results. Two mathematicians of the 19th century, George Boole (1815–1864) and Gottlob Frege (1848–1925), independently came up with and expanded upon Leibniz's ideas.

Hence, it is appropriate to consider Leibniz, Boole and Frege to be the pioneers of modern logic. The next three sections cover them in detail.

Leibniz: Anticipation of Modern Logic

Leibniz not only realized that Aristotle's system of logic had considerable shortcomings, but also saw that these defects could not be addressed by a few changes. So he set out to create a new foundation of logic.

The effort was successful, but here and there his construction contained gaps that apparently kept him from publishing the results. As a consequence, his work only became known 150 years later, in the middle of the 19th century. There are two ways to evaluate Leibniz's work on logic.

First, one may examine his results, locate the gaps, fill them, and then compare the now-complete results with modern logic. When this is done,[288] the conclusion emerges that Leibniz anticipated many logic results of the 19th and 20th century.[289]

The second way of evaluating Leibniz's results does not try to fill gaps using our knowledge of modern logic.[290] Thus, no attempt is made to modify the results found in Leibniz's notes.

This alternate way of looking at Leibniz's work results in the same conclusion: His construction is an ingenious anticipation of modern logic. But it also leads to the conclusion that the groundbreaking work on modern logic done by Boole and Frege in the 19th century—covered in the next two sections—was done independently of Leibniz's results.

Leibniz also envisioned a *calculus ratiocinator*[291] (computing evaluator) that, depending on interpretation and viewpoint, anticipates either a formal inference engine, a computer program, or a computer itself. In the third case, the calculator designed by Leibniz and described in Chapter 7 can be viewed as the first step in the direction of a computer.

The concept of calculus ratiocinator is closely connected with Leibniz's idea of a precise universal language he termed *characteristica universalis*[292] (universal characteristics). It was to support a precise encoding of the facts of the world. The calculus ratiocinator then would carry out an evaluation of these facts and verification of conclusions.

Thus, Leibniz anticipated the key idea of *artificial intelligence*[293] formulated in the 20th century. Indeed, he must be considered the founder of that discipline.

The above discussion has glossed over the variance among modern interpretations of the calculus ratiocinator and the characteristica universalis.[294] Nevertheless, there is universal agreement that Leibniz was a visionary for logic and its uses in mathematics and the everyday world.

We jump forward 150 years to Boole and Frege, who were not aware of Leibniz's results when they started their efforts. We first discuss Boole. His work marks the beginning of the 100 years of proof developments mentioned in the introduction of this chapter.

Boole: Reliable Logic Computation

Boole's work was motivated by the work of Augustus De Morgan (1806–1871), who defined basic concepts of logic such as propositions and rules of deduction.[295]

But Boole's approach was different from that of De Morgan, or for that matter, of Leibniz. While the latter

Augustus De Morgan.[296]

mathematicians strove for elaborate ways to formulate facts and relationships in logic statements, Boole aimed for a simple enough framework that allowed computations.

Indeed, Boole's most impressive achievement is the first complete calculus where mathematical operations reliably solve logic problems.[297]

AN INVESTIGATION

OF

THE LAWS OF THOUGHT,

ON WHICH ARE FOUNDED

THE MATHEMATICAL THEORIES OF LOGIC
AND PROBABILITIES.

BY

GEORGE BOOLE, LL.D.

PROFESSOR OF MATHEMATICS IN QUEEN'S COLLEGE, CORK.

LONDON:
WALTON AND MABERLY,
UPPER GOWER-STREET, AND IVY-LANE, PATERNOSTER-ROW.
CAMBRIDGE: MACMILLAN AND CO.
1854.

Left: George Boole.[298]
Right: Boole's The Laws of Thought, 1854[299]

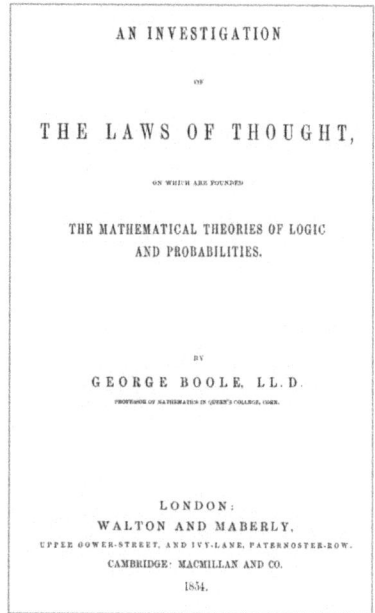

We couldn't state the key idea underlying the computational process more clearly and concisely than done by Boole in his book *The Laws of Thought:*[300]

"But as the formal processes of reasoning depend only upon the laws of the symbols, and not upon the nature of their interpretation, we are permitted to treat the above symbols x, y, z [which represent classes of objects that do or do not have certain properties] as if they were quantitative symbols of the kind above described.

"We may in fact lay aside the logical interpretation of the symbols in the given equations, convert them into quantitative symbols, susceptible only

*of the values 0 and 1; perform upon them as such all the requisite pro-
cesses of solution; and finally restore to them their logical interpretation."*
[emphasis in the original]

Subsequently, Boole's system was simplified.[301] The resulting system is now called *Boolean algebra.*[302] Its formulas have an equivalent
expression in the sentences of *propositional logic.*[303]

Frege: Definitions for Logic

In the groundbreaking book *Begriffsschrift*[304] (Treatise of Concepts),
Frege created definitions of clarity and precision for logic that are
still in use today.[305] Translated into English, the subtitle summarizes the key idea of the book as follows: *A language of formulas of
pure thought, inspired by arithmetic.*

Left: Gottlob Frege.[306]
Right: Frege's Begriffsschrift, 1879.[307]

For the first time, Frege's definitions allowed rigorous modeling
of the logic structure of complex situations, whether occurring
in mathematics or the world. Fourteen years later, in 1893, Frege

dared to construct in *Grundgesetze der Arithmetik*[308] (Fundamental Laws of Arithmetic) a foundation of arithmetic relying solely on logic. In-between, he had discussed a number of philosophical questions about arithmetic in a related book[309] where, in his words,[310] "[he] had tried to show that the arithmetic is a branch of logic which need not base the motivation for proofs on any experience or image."

In Grundgesetze der Arithmetik, this claim was to be proved by showing that the simplest laws of arithmetic could be established with just the tools of mathematical logic of the Begriffsschrift. The book represents a monumental effort where hundreds of logic diagrams prove the claims.

— 76

GRUNDGESETZE

DER ARITHMETIK.

Begriffsschriftlich abgeleitet

von

Dr. G. FREGE
A. O. PROFESSOR AN DER UNIVERSITÄT JENA.

I. Band.

JENA
Verlag von Hermann Pohle
1893.

§ 56. *Zerlegung.*

Um nun den Satz

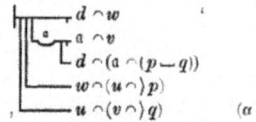

$$d \frown w$$
$$a \frown v$$
$$d \frown (a \frown (p - q))$$
$$w \frown (u \frown) p)$$
$$u \frown (v \frown) q) \qquad (\alpha$$

(§ 54, π) zu beweisen, müssen wir auf ($\mathcal{A}$) zurückgehen. Daraus leiten wir den Satz

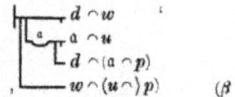

$$d \frown w$$
$$a \frown u$$
$$d \frown (a \frown p)$$
$$w \frown (u \frown) p) \qquad (\beta$$

ab. Um von diesem aus (α) zu erreichen, müssen wir den Satz

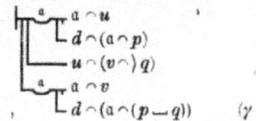

$$a \frown u$$
$$d \frown (a \frown p)$$
$$u \frown (v \frown) q)$$
$$a \frown v$$
$$d \frown (a \frown (p - q)) \qquad (\gamma$$

haben, der nach Regel (5) hervorgeht aus

Left: Frege's Grundgesetze der Arithmetik, 1893[311]
Right: Logic diagrams of Grundgesetze der Arithmetik, detail of p. 76 left column.

As Frege was completing the book, he received a letter from Bertrand Russell (1872–1970) that pointed out a devastating inconsistency in a basic assumption of the book. Frege's response was a painstaking analysis in a 13-page appendix to discover the root of the difficulty and suggest possible remedies. Unfortunately, at the end he had to admit that he did not have a satisfactory solution.[312]

Bertrand Russell.[313]

The problem concerned Frege's definition of a *set*, which generally is a collection of *elements*. For example, we may declare *primes* to be the set whose elements are the prime whole numbers.

The concept of set is essential when one constructs any area of mathematics using logic: Mathematical claims can then be rephrased as logic statements claiming certain elements to be in a set, or a certain set to be contained in another set, and so on.

Frege defined a set to be any collection that can be defined by a logic condition. For example, the set of prime numbers n can be defined by imposing the logic condition that n must be a natural number that cannot be factored in a nontrivial way. Unfortunately, Frege's definition admits sets with contradictory properties,[314] as discovered by Russell.

We move forward to the beginning of the 20th century. At that time, a historic battle within mathematics started about the validity of mathematical assumptions and proof techniques.

We saw an early skirmish of that battle in Chapter 4, where Cantor defended his revolutionary ideas about infinity against Kronecker's demand of finitism. But now a full-blown battle unfolded over choice in assumptions and proofs.

The story begins with David Hilbert (1862–1943), who arguably was the most prominent mathematician of the late 19th and early 20th century. His interests were universal, and his results had far-reaching impact.[315] Here, we will focus on a set of problems he posed right at the beginning of the 20th century.

Hilbert: 23 Problems

In 1900, Hilbert posed 23 unsolved problems that were of eminent importance. Indeed, they motivated a large part of the mathematical research of the 20th century.[316]

Among them was Cantor's continuum hypothesis, which says that there is no set whose cardinality lies strictly between that of the natural numbers and the real numbers.

Put differently, it claims that the real numbers constitute a smallest uncountable set.

David Hilbert.[317]

As part of that hypothesis, Hilbert asked that a certain result called the *well-ordering theorem* be proved, where *well ordering* is defined as follows: A set with a given ordering of the elements is *well ordered* if every nonempty subset has a smallest element.[318]

The real numbers with their natural ordering are not well ordered. For example, take the subset of real numbers greater than 0. It does not have a smallest number, since for any $x > 0$ of the set, $\frac{x}{2}$ is also in the set. Thus, there cannot be a smallest $x > 0$.

The *well-ordering theorem* says that, for every set, there exists some ordering such that the set becomes well ordered. In particular, the theorem implies that there is an ordering of the set of real numbers such that it becomes well ordered, a strange claim at the time since nobody had a clue how to do this.

Zermelo: Axiom of Choice

In 1904, Ernst Zermelo (1871–1953) set out to prove the well-ordering theorem. For this task, he invented an axiom that later was called the *axiom of choice*. At the time, it looked to be a totally self-evident axiom, and Zermelo had no compunction in defining and using it. We describe the axiom next.

Imagine that a set is represented by a basket with a lid. Any elements of the set are represented by items in the basket. Thus, an empty basket corresponds to the empty set.

Consider a long row of baskets of the above kind. Somebody assures us that each basket contains at least one item. We are asked to get exactly one item from each basket and set it down next to the basket, as sort of demonstration that the basket indeed isn't empty.

Ernst Zermelo.[319]

Basket with lid.[320]

This is easily done. We step up to each basket, open the lid, grab an item, and put it down next to the basket. The task may be time-consuming if the row of baskets is long, but in principle we can do this.

Now make the row of baskets longer and longer. At the same time, imagine the baskets to be labeled, say with the natural numbers 1, 2, 3, . . . , with no limit.

We are still assured that each basket is nonempty. We are asked to carry out the above process, but with the additional condition that we are to assign the label of each basket to the removed item.

Thus we get item$_1$, item$_2$, item$_3$, ... Of course, we cannot complete the process in finite time. But we can imagine doing it forever, and in the process will achieve the goal.

We come to an even bigger collection of baskets. Each basket has again a label, but there are so many baskets that we cannot arrange them in a row. Instead, there is an arbitrarily large collection L of labels, and for each label x in L, denoted by $x \in L$, there is exactly one basket labeled with x.

Once more, we are told that each basket is nonempty, and we are asked to remove one item from each basket and set it down next to the basket. We can readily do this for a given basket; say, we remove item$_x$ from basket x.

But we are hard-pressed to describe a removal process that for any collection L of indices will complete the task even in infinite time.

The *axiom of choice* states that the desired removal is not only possible, but can be carried out in one step regardless of the size of the collection L.

So given nonempty baskets labeled by L, in one fell swoop we obtain item$_x$ from basket x, for each $x \in L$. For the case of countable L—for example, in the above case of labels 1, 2, 3, . . .—the axiom becomes the less demanding *axiom of countable choice.*[321]

The axiom of choice had been implicitly employed prior to Zermelo. For example, Cantor used it in the construction of certain infinite sets without recognizing that he was invoking something outside standard set theory.

Zermelo was first to realize, in 1904, that the axiom needed explicit formulation. To him, the axiom of choice seemed of self-evident reasonableness and simplicity. Accordingly, he felt free to use it in his proof of the well-ordering theorem.

In 1908, Zermelo proposed axioms of set theory that were later modified by Abraham Fraenkel (1891–1965). The resulting *Zermelo-Fraenkel set theory* included the axiom of choice. That system is denoted by *ZFC*, where the

Adolf Abraham Halevi Fraenkel.[322]

"C" signifies use of the axiom of choice. It was soon determined that the axiom of choice had profound consequences. Accordingly, Zermelo-Fraenkel set theory without the axiom of choice became of interest; it is denoted by ZF.

Banach and Tarski: Paradox

In 1924, Stefan Banach (1892–1945) and Alfred Tarski (1901–1983) proved the following astonishing result, now called the *Banach-Tarski paradox:*[323]

The axiom of choice of ZFC may be used to divide a solid 3-dimensional sphere into a great many small pieces, and then reassemble these very pieces and obtain two solid 3-dimensional

Banach-Tarski Paradox: A solid ball is turned into two solid balls of the same size.[324]

spheres, each with the *same* diameter as the original one. In short, the axiom allows a doubling of material.

Left: Stefan Banach.[325]
Right: Alfred Tarski.[326]

The paradox proved that the axiom of choice has strange and counterintuitive effects, and thus destroyed the notion that the axiom could be justified by appeal to physical everyday processes such as removing items from baskets.

Six years earlier, in 1918, L. E. J. Brouwer (1881–1966) had raised a more fundamental objection to proof processes based just on ZF, and thus had started a new philosophy of mathematics called *intuitionism*.[327]

At that time, Brouwer was known for several profound results in *topology*, the branch of mathematics concerned with properties of space that are independent of stretching and bending.[329]

L. E. J. Brouwer.[328]

An example is his *fixed-point theorem*,[330] which stands out in a long list of fixed-point results due to its widespread use in various fields of mathematics.

An illustrative example application[331] of the theorem is as follows. Take two sheets of paper of same size. Let one page cover the other one. Then each point of the bottom page corresponds to a unique point of the top page.

Now crumple up the top page in any way and position it so that it does not extend beyond the boundary of the bottom page. Then there is a point of the crumpled page such that the corresponding point of the flat page lies vertically below.

Brouwer: Intuitionism

Brouwer's intuitionism[332] views mathematics as a constructive process. In particular, a mathematical object can be claimed to exist only if it can be constructed.

As an example, suppose we have a set S that we guess to be not empty. We wish to remove one element from S for further mathematical manipulation. Unfortunately, we somehow fail to assemble direct logic arguments that provide the desired element.

At this point, the axiomatic method allows an indirect path to our goal. We assume that S is empty and, with appropriate arguments of logic, aim for a contradictory result that invalidates that assumption.

Suppose we succeed in this endeavor. Thus, we have shown that the statement "S is empty" is false. Using Aristotle's *law of the excluded middle*,[333] there is only one alternative, which means that the statement "S is not empty" is true.

Thus, S contains at least one element. We arbitrarily pick one such element and use it in further mathematical manipulation, as desired.

The intuitionist goes along with this method right up to the point where the assumption "S is empty" results in a contradiction.

But the next step—claiming that S is not empty and extracting one item from S—is not permissible. Indeed, the intuitionist obtains such an item only by a suitable construction and not simply by the fact that the claim "S is empty" has produced a contradiction.[334]

In 1921, Hermann Weyl (1885–1955) clarified and expanded Brouwer's ideas.[335]

Weyl eventually became one of the most influential mathematicians of the 20th century. He achieved outstanding results in several branches of mathematics including number theory, and in theoretical physics.

During the second half of the 1920s,

Hermann Weyl.[336]

Weyl began to realize that the restrictions imposed by intuitionism would cripple mathematics. At that same time, Hilbert and Brouwer engaged in a monumental struggle about the future direction of mathematics. The war culminated in an ugly battle about membership on the editorial board of the leading mathematical journal *Mathematische Annalen.*[337]

On one hand, Hilbert wanted to ensure that the editorial board accepted the axiomatic method, in which rules of logic derive results from basic axioms.

These rules allow simple steps such as claiming use of an element of a set that has been proved to be nonempty by contradiction—see the earlier example—as well as much more complicated creation and manipulation of functions or sets based on the axiom of choice.

In fact, by the end of the 1920s, the axiom of choice was fully accepted as part of Zermelo-Fraenkel set theory, despite disturbing results such as the Banach-Tarski paradox. As stated earlier, that version is denoted by *ZFC*, and the version without the axiom of choice by *ZF*. Today, the system *ZFC* is considered the foundation for most of mathematics.[338]

On the other hand, Brouwer wanted to see all such steps ruled out unless supported by constructive arguments. Acceptance of that severe condition would have invalidated many results of Hilbert and others, and would have precipitated a historic destruction of mathematics.

No wonder Hilbert and like-minded mathematicians waged a ferocious battle against Brouwer's demand that mathematics be based on the concepts of intuitionism.

Fortunately for mathematics, Hilbert's side won, and today nobody seriously questions whether non-constructive results are acceptable.

Looking back now, almost 100 years later, Brouwer's life is seen to be a tragedy: Here was a brilliant mathematician who created groundbreaking results in topology during the four years spanning 1909–1913 and thus had started an illustrious mathematical career.

Hilbert was so impressed that in 1912 he helped Brouwer obtain a regular academic appointment at the University of Amsterdam.[339]

With financial support secured, Brouwer embarked on development of the philosophy of intuitionism to the exclusion of everything else. This effort spanned decades of his life, but eventually turned out to be futile.

We close the discussion of intuitionism with a quote by Weyl, who by 1949 had become disillusioned with intuitionism. His statement rejects intuitionism, but also acknowledges Brouwer's lofty goals:[340]

"Mathematics with Brouwer gains its highest intuitive clarity. He succeeds in developing the beginnings of analysis in a natural manner, all the time preserving the contact with intuition much more closely than had been done before.

"It cannot be denied, however, that in advancing to higher and more general theories the inapplicability of the simple laws of classical logic eventually results in an almost unbearable awkwardness. And the mathematician watches with pain the greater part of his towering edifice which he believed to be built of concrete blocks dissolve into mist before his eyes."

We go back to the beginning of the 1910s, when Alfred North Whitehead (1861–1947) and Russell proposed a novel construction of mathematics based on logic.

In 1893, Frege had attempted such a construction in the Grundgesetze der Arithmetik, but had failed due to a contradictory definition of sets.

Whitehead and Russell: Principia Mathematica

During the period 1910-1913, Whitehead and Russell published the *Principia Mathematica*, a landmark achievement that in three volumes defined a foundation of mathematics.[341]

Their approach was based on the long-held belief that all of mathematics could be constructed from elementary principles: One just had to find the right way to define the fundamental concepts, determine a reliable construction method, and then all of mathematics could be built.

For Whitehead and Russell it seemed obvious, just as it did for Frege, that these fundamental principles were to be found in the logic created in the 19th century.

PRINCIPIA MATHEMATICA

BY

ALFRED NORTH WHITEHEAD, Sc.D., F.R.S.
Fellow and late Lecturer of Trinity College, Cambridge

AND

BERTRAND RUSSELL, M.A., F.R.S.
Lecturer and late Fellow of Trinity College, Cambridge

VOLUME III

Cambridge
at the University Press
1913

Above: Alfred North Whitehead.[342]
Right: Principia Mathematica, Vol. III, 1913.[343]

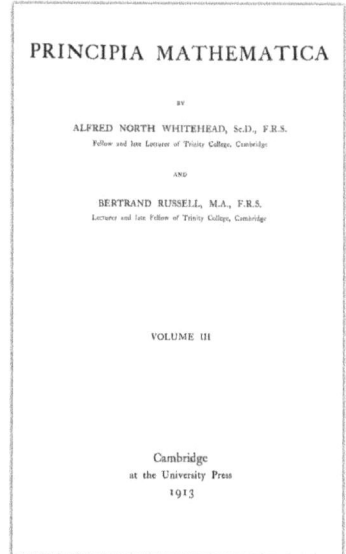

Whitehead and Russell worked carefully to avoid the conundrum of Frege, where the definition of sets had introduced a fatal contradiction.

The painstaking approach started with the basic rules of propositional logic and then built up result after result through an intricate network of logic formulas.[344] Part of the construction was a complicated hierarchy of sets.

After 379 pages of volume I, a logic statement equivalent to the arithmetic result $1 + 1 = 2$ was proved. Due to that slow pace, the three volumes covered only set theory, cardinal numbers, ordinal numbers, and real numbers.

But it seemed evident that the construction method, suitably continued, would produce a large portion if not all of mathematics.

A second edition was published in 1925, and an abbreviated version in 1997, showing that interest in the work continued over decades.

Comments ranged from suggestions for modifications to fundamental criticism. We will skip that extensive material[345] and instead focus on two aspects: the *completeness* and *consistency* of systems of axioms. The definitions of these two terms are as follows:

A system of axioms is *complete* if every true statement can be proved in the system. It is *consistent* if it does not contain a contradiction.[346]

The *Principia Mathematica* starts with propositional logic, which is complete and consistent. At the time, it was assumed that the two properties would be maintained by suitable construction of results.

That aspect, indeed, the general problem of completeness and consistency of mathematics, was taken up by Hilbert. He wanted to place all of mathematics on a reliable foundation where completeness and consistency were proved throughout.[347]

Hilbert: Completeness and Consistency

In 1880, the physician and physiologist Emil du Bois-Reymond (1818–1896) argued in a speech before the Berlin Academy of Sciences that the Latin maxim *ignoramus et ignorabimus* (we do not know and we will not know) applies to certain parts of the world.[349]

Emil du Bois-Reymond.[348]

In particular, mankind would never understand the nature of matter and force, the origin of motion, and the origin of simple sensations. Bold predictions are often wrong. This certainly applies here: Matter, force, and motion have been explained by Einstein's theory of relativity and quantum physics; and simple sensations have been investigated by brain science, along with many other interactions of mind and body with the external world.

In 1930, Hilbert delivered a celebrated address to the Society of German Scientists and Physicians where he argued passionately against the mentality of *ignorabimus:*[350]

"We must not believe those, who today, with philosophical bearing and deliberative tone, prophesy the fall of culture and accept the *ignorabimus*. For us there is no *ignorabimus*, and in my opinion none whatever in natural science.

"In opposition to the foolish *ignorabimus* our slogan shall be: Wir müssen wissen – wir werden wissen! (We must know – we will know!)"

Part of Hilbert's tombstone: Wir müssen wissen – wir werden wissen (We must know, we will know).[351]

The famous statement "Wir müssen wissen – wir werden wissen" eventually was engraved on Hilbert's tombstone.

Hilbert's arguments against Du Bois-Reymond's predictions were well justified. Hadn't several problems open since antiquity been solved in the 19th century? Didn't Frege design the Begriffsschrift, which allowed accurate encoding of all facts in logic? Weren't the axioms of Zermelo-Fraenkel a precise formulation of set theory? Didn't Whitehead and Russell show that basic concepts of numbers and related operations could be developed just starting with logic?

So Hilbert decided that there was the opportunity—indeed the duty—to answer once and for all the fundamental questions of mathematics. In particular, the nagging uncertainty about completeness and consistency of systems, some used for thousands of years, had to be removed.

In the 1920s, and thus before the talk cited above, Hilbert had already started a project, now called *Hilbert's Program*, that was to create a secure foundation for all of mathematics.

His program had a number of goals.[352] The two most important ones were proof of *completeness* and *consistency* of all axiomatic systems.

The process proving completeness and consistency of mathematical systems obviously would use mathematical arguments. How could Hilbert be sure that these arguments themselves were not flawed?

To preclude that disastrous possibility, he outlined a very restricted mathematics called *finitary* that unquestionably could be used to prove completeness and consistency for general mathematical systems.[353]

With that finitary proof machinery established, he and others set out to process the existing mathematical systems one by one, each time searching for proofs of the desired features. Early successes for simple mathematical systems supported Hilbert's optimism that the program could indeed be carried out. For example, the theory of only addition of natural numbers and of multiplication of the positive integers was proved to be complete and consistent.[354]

But then progress came to a halt: In 1931, Kurt Gödel (1906–1978), at age 25, published devastating results that limited what could *ever* be ascertained by mathematical arguments. That work and other results in logic prove Gödel to be one of the greatest logicians of all time.

Kurt Gödel, ca. 1926.[355]

Gödel: Incompleteness

Gödel investigated what could ever be established for a given axiomatic system when all arguments are based just on those axioms and do not use any external assumptions.

That effort produced the following two results in 1931, now called *incompleteness theorems*, that forever limit what can be proved that way:[356]

First incompleteness theorem: For any formal mathematical system that contains a certain amount of elementary arithmetic, there are statements composed in the language of the system that can be neither proved nor disproved in that system.

Example systems are Zermelo-Fraenkel set theory without axiom of choice (*ZF*) or with that axiom (*ZFC*). For either system, there are statements formulated in the language of that system that cannot be proved or disproved. If such statements are added as axioms, then there are new statements that cannot be proved or disproved. Thus, there is no remedy of incompleteness of the system.

Second incompleteness theorem: For any consistent system that contains a certain amount of elementary arithmetic,[357] consistency of the system cannot be proved in that system.

This second result also applies to *ZF* and *ZFC*. That is, one cannot possibly prove consistency in those systems.

Taken together, the two results dealt a fatal blow to Hilbert's program:[358] There was no more hope that any mathematical system of reasonable sophistication could be proved complete or consistent, just using the axioms of the system itself.

In particular, for *ZFC* there could never be a proof of consistency constructed within that system. Yet, that theory was, and still is, the foundation of most of mathematics.

So after centuries of construction, mathematicians have created a foundation for almost all of mathematics that might be faulty. This is a mind-boggling conclusion.

Now *ZFC* has been used almost 100 years, and no case of inconsistency has surfaced. If we were to argue like Wallis of the 17th century,[359] we would say that there have been plenty of experiments to support the belief that there is no inconsistency.

That sort of argument for validity was rejected when Wallis used it to justify his results. Yet exactly this argument is made today regarding the most fundamental part of mathematics.

There are two alternatives that avoid the consistency problem at the foundation of mathematics.

First, we could proceed like physicists, who still use Newton's model of the world instead of Einstein's theory of relativity or quantum physics initiated by Max Planck (1858–1947), as long as the errors are so small as to be irrelevant.

Analogously, we could have a construction of mathematics guided by the philosophy of Wallis where all results are verified by experiments and used until a serious error surfaces. But what a confusing world it would

Max Planck.[360]

be! So no mathematician would agree to such a course of action.

Second, we could confine mathematics to finiteness, as proposed by Kronecker, and then handle complicated concepts like $\sqrt{2}$ and π by approximations.

Precisely this restriction to finiteness is happening in modern computation, almost without exception.

But a general restriction to finiteness would remove soaring ideas such as Cantor's construction of the infinite cardinal and ordinal numbers. Indeed, it would reduce mathematics to a drab world.

So, mathematicians have accepted a compromise. They tolerate the potential for inconsistency in the foundation of mathematics, but from then on carry out a construction with unassailable proofs and thus can build result upon result.

The conclusions are *relatively valid*: If the foundation is consistent, then all results hold. If an inconsistency surfaces in the foundation, mathematicians hope—indeed anticipate—that the problem can somehow be remedied by a suitable modification that does not destroy the entire edifice. They also accept that the method cannot establish all true results as indeed true, an unavoidable shortcoming established by Gödel's first incompleteness theorem.

For the next step in the history of mathematical proof, we need the concept of *independence of axioms*, defined as follows.

Let S be a consistent set of axioms, and define A to be some other axiom. Then axiom A is declared to be *independent* from S if the following two systems are consistent: S with A added, and S with the negation of A added.[361]

An important question since the definition of ZF was: Assuming that the axioms of ZF are consistent, are the axiom of choice and the continuum hypothesis independent of ZF?

Gödel partly proved this. In 1938, he showed the following: If ZF is consistent, then ZF with the axiom of choice added is consistent as well. In 1940, he proved the analogous result for the continuum hypothesis.[362]

But then no further progress was made. Indeed, many others tried to answer the open question about consistency when the negation of the axiom of choice or of the continuum hypothesis is added.

The problem seemed so difficult that young mathematicians were counseled not to work on that seemingly hopeless problem. Fortunately, one mathematician ignored that advice.

Cohen: Independence of Choice

In 1963, at age 29, Paul J. Cohen (1934–2007) published the following result:[363]

If ZF is consistent, then addition of the negation of the axiom of choice or of the negation of the continuum hypothesis results in another consistent system.

That proof plus Gödel's earlier conclusion established the long-sought
result that the axiom of choice and
the continuum hypothesis are independent of ZF.

Gödel sent a comment to Cohen, a
draft of which has survived. It says,

"Let me repeat that it is really a
delight to read your proof of the
ind[ependence] of the cont[inuum]
hyp[othesis]. I think that in all essen-

Paul J. Cohen.[364]

tial respects you have given the best possible proof & this does not
happen frequently. Reading your proof had a similarly pleasant
effect on me as seeing a really good play."[365]

Cohen's proof was based on *forcing*,[366] a method he invented for
proving consistency and independence of results. The method has
become a standard tool for work on the foundation of mathematics.

Using forcing, he also showed that the independence result for
the continuum hypothesis remains correct when the larger system
ZFC, which is ZF plus the axiom of choice, is used.

That is, if ZFC is consistent—which is the case if ZF is consistent—
then the continuum hypothesis is independent of ZFC.

We thus can declare either ZFC with the continuum hypothesis
added or with the negation of that hypothesis added to be our
foundation of mathematics, and are assured that either choice is
consistent if ZF is consistent. Which of the two possibilities should
be selected for the mathematics of the future? There is substantial
debate about this, with no easy resolution in sight.[367]

That debate once more shows that development of the foundation
of mathematics is an ever ongoing process.

In all informal discussions about mathematics, we have used the
image of a building where the load-bearing columns in the base-

ment correspond to the foundation of mathematics. A different and richer image was proposed in 1948 by Nicolas Bourbaki;[368] the name is a collective pseudonym of a group of mathematicians who have attempted a comprehensive coverage of mathematics:

"Mathematics is similar to a large city where suburbs grow into the surrounding land and where the center is periodically rebuilt, each time according to a clearly defined plan and according to a new, more impressive order."[369]

But even this expanded picture does not capture the richness of mathematical developments. In the terminology of the Bourbaki paper, we encounter next the construction of an extraordinary bridge connecting two distant suburbs of the city of mathematics.

Wiles: Proof of Fermat's Last Theorem

In 1637, Fermat claimed the following, now known as *Fermat's last theorem*:[370]

The equation $x^n + y^n = z^n$, with n any natural number greater than 2, has no solution where x, y, and z are natural numbers.

For the case $n = 2$, there are of course solutions with natural numbers; a collection of such solutions, now called Pythagorean triples, was already created in ancient Babylon, as cited earlier in this chapter.

Fermat wrote in the margins of one of his books that he had a proof.[371] But it is most likely that he was mistaken in this assessment, given the complexity of the eventual proof.

Over centuries, Fermat's claim was proved[372] for ever larger values of n. Computers of the 20th century allowed a massive numerical evaluation that confirmed Fermat's claim to be correct up to $n = 4,000,000$.

In 1995, more than 350 years after Fermat posted the claim, Andrew John Wiles (1953–) published a proof of the theorem.[373]

It was a stunning achievement. The seeds for the proof were planted in 1955, when Yutaka Taniyama (1927–1958) and Goro Shimura (1930–) created an utterly daring conjecture: Two mathematical areas that to all appearances were completely different, were claimed to be fundamentally linked.[374]

The proposed link seemed preposterous. In terms of Bourbaki's city with suburbs, the conjecture envisioned a bridge connecting two far-apart suburbs. That conjecture, unlikely as it seemed, turned out to be correct

Andrew John Wiles.[375]

and is now called the *modularity theorem*.[376] Wiles proved a special version of the conjecture that was sufficient to establish Fermat's last theorem.

The story is too complex to be covered here, but has been well described.[377] Building on Wiles's work, other mathematicians proved the remaining portion of the modularity theorem during 1996–2001.

Summary

The chapter has traced the history of the concept of mathematical proof, beginning with Babylonian results based on geometry and groundbreaking efforts of the ancient Greeks.

The next milestone was Leibniz, who anticipated the concepts of modern logic so essential for reliable proofs. Boole and Frege then created key parts of that machinery.

Late in the 19th century and continuing into the 20th century, mathematicians tried to create a precise foundation of mathematics and reliable methods for proofs.

That process involved Frege, Zermelo, Fraenkel, Whitehead, and Russell. It essentially ended up with Zermelo-Fraenkel set theory as the basis for most of mathematics.

A struggle about the use of axioms and the interpretation of mathematical steps pitted Brouwer and Weyl, the proponents of intuitionism, against mathematicians led by Hilbert.

The latter group freely used axioms to build astonishing results. Hilbert also launched a drive to put all of mathematics on a permanent, unassailable foundation.

That effort collapsed when Gödel proved that this goal will forever be unattainable. In particular, consistency of the foundation axioms will always be in doubt. Gödel also showed that two advanced axioms—the axiom of choice and the continuum hypothesis—can be added to ZF without fear of inconsistency, assuming that ZF itself is consistent.

Cohen proved the very difficult additional result that the negated versions of these axioms can also be added without loss of consistency.

We thus have wide-open choices about the foundation of mathematics, where the axiom of choice or the continuum hypothesis can be added to ZF or declared to be not valid. All such choices are allowed under the assumption that ZF is consistent.

At this time, the addition of the axiom of choice is fully accepted, and the resulting system ZFC is generally used. But addition of the continuum hypothesis or its negation is still being debated.

Lastly, Wiles's proof of Fermat's last theorem shows that seemingly disparate areas of mathematics are sometimes closely connected. This result has motivated the current search for profound links joining widely different regions of mathematics.

The next chapter looks at an area closely related to mathematics: computing machines. The need for such machines existed for several thousand years, but first steps toward their creation were taken less than 400 years ago.

Ingenious ideas then produced mechanical calculators for the four basic arithmetic operations and, eventually in the 19th century, the first mechanical computer.

All such equipment was quite complex, in large part due to the encoding of numbers: The decimal system was used—a seemingly natural choice given its proven effectiveness for manual computation.

But that notion was simply mistaken, and the first computer designed in the 19th century with that encoding was never built due to its extraordinary complexity.[378]

Then in the 1930s, one person shifted to binary encoding, developed the needed mathematical machinery, and in a six-year period built the first programmable computer. It was an incredible achievement.

Thus, the history of computation is yet another demonstration that the development of mathematics, or of related equipment as described in the next chapter, isn't a straightforward progression of ever-better results, but can for centuries proceed at a modest pace and then jump forward in one gigantic leap.

7

Computing Machines

When mathematics consisted just of addition and subtraction of natural numbers, computation was readily accomplished with pebbles, sticks, or other small items.

Abacus.[380]

Piles of items represented the numbers. Two piles were combined to carry out addition. Portions of piles were removed for subtraction. Various versions of the *abacus*[379] carry out these steps by moving beads. Computations then became more complicated due to two developments.

First, a proliferation of mathematical operations such as multiplication, division, taking roots, and exponentiation made manual computation ever more tedious.

Second, human ingenuity came up with increasingly sophisticated models of various aspects of the world, such as the movement of the planets. As these models became more complicated, the computational tasks quickly increased in complexity to the point where manual evaluation became very difficult.

Thus, there were two problems: The handling of sophisticated mathematical operations, and the manipulation of complicated models.

Initially, mathematicians addressed these problems in two independent developments.

First, they reduced complicated mathematical operations to simpler ones and built computing machines that exploited these results.

Second, they represented mathematical models of the world with physical models whose operation directly gave the desired answers. The latter approach was constrained since available materials and technology limited what could be built.

Indeed, as computing machines became more powerful, mathematicians gradually abandoned the construction of physical models and instead evaluated mathematical models of the world directly.

The development of computing machines proceeded at a much slower pace than that of mathematical theory. One might argue that the slow rate of progress was largely due to the fact that, up to the 19th century, few suitable materials were available for the construction of computing machines, craftsmanship was limited by imprecise machining equipment, and construction cost was huge when compared with the cost of everyday living.

To some extent, that reasoning is correct. But a more incisive explanation for the slow progress is that mathematicians looked at the problem of computation the wrong way.

They thought that machine computation should be done in the decimal system since it had proved to be effective for hand computation. That viewpoint persisted until the 1930s.

At that time, a major shift occurred: It was recognized that computation in the binary system was not only more efficient, but could be carried out with much simpler machines.

The new viewpoint led to rapid development and construction of powerful computing equipment starting in the 1930s and continuing today at an ever increasing pace.

This chapter covers milestones of that history, beginning with early developments and ideas, and ending just before the invention of the electronic computer.

We restrict ourselves to these milestones for two reasons. First, they demonstrate how the human mind was able to convert abstract concepts of mathematics into impressive physical machinery when few materials were available and tools for production were rather primitive.

Second, the subsequent milestones of the electronic age are so numerous and complex that discussion within a book chapter would only be superficial.

Six Milestones

Computation in the decimal system demands reliable transfer of the values 1 to 9 from one counter to another. Gears are the perfect means for that transfer. The first milestone is an impressive use of gears in a physical model constructed in ancient Greece.

Addition requires a mechanism for the carry operation; for example, in the computation of $45 + 37$, the lowest digits 5 and 7 are added, resulting in 12. The 2 of 12 is recorded as the last digit of the eventual sum, and the 1 is carried over for addition to the second digits 4 and 3.

Accommodation of a single carry step in a mechanical calculator is not so difficult. But the process becomes more complicated when a carry step triggers a cascade of additional cases; for example, when 1 is added to $9,999$. The second milestone is the invention of the first machine with reliable carry operation regardless of the length of such cascades.

It is attractive to represent multiplication by repeated addition, and division by repeated subtraction. Automated execution of these processes requires storage and retrieval of multiplier and divisor values. This is readily accomplished with gears if the number of cogs of a given gear can be adjusted stepwise from a maximum of 9 to a minimum of 0. One full rotation of such a gear then transfers the number of cogs as the value. It seems nearly impossible to design such a gear. Invention of not just one but two ingenious gears with that feature is the third milestone.

Now consider sequences where the four basic operations are to be done repeatedly according to some rules. How can this process be accomplished with gears? The invention of a sophisticated machine with gears for this task constitutes the fourth milestone. Due to its complexity and huge construction cost, the machine was never built.

For centuries, it was always assumed that the values of any function could be computed. The fifth milestone is the invention of a simple computing machine with which that notion was proved to be incorrect. In fact, that machine set a standard of computability.

The sixth and last milestone is a new way of thinking about computation. The new approach relies on the earlier-mentioned shift to the binary system. The theory of such computation was developed and implemented in less than six years, resulting in the world's first programmable, fully automatic computer.

Let's start with the first milestone. It marks the use of gears for reliable transfer of rotation in a sophisticated physical model.

Antikythera Mechanism: Gears

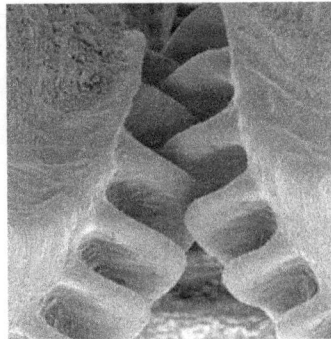

Gear mechanism discovered in *Issus coleoptratus*, a planthopper species common in Europe.[381]

Gears were used in China in the 4th century BCE to transfer rotation from one shaft to another in a fixed ratio.[382] Evolution developed that concept as well: The juvenile form of a planthopper insect found in many European gardens, species *Issus coleoptratus*, has gear-like serrations in its hind legs that force coordinated movement during jumps. The plant-hopper sheds the gears when it becomes an adult.[383]

From the 4th century BCE onward, gears were used in various physical models representing aspects of the real world. An outstanding example is the *Antikythera mechanism*[384] found in 1901-1902 in a shipwreck off the Greek island of Antikythera.

Antikythera mechanism, reconstruction[385]

The mechanism was created in Greece during the 2nd century BCE for predicting astronomical positions and eclipses as well as cycles of the ancient Olympiads.

The output was computed by a complex arrangement of bronze gears[386] housed in a 13.4 x 7 x 3.5 in. wooden box. The model was operated with a small hand crank, now lost. The configuration of gears was reconstructed from 82 badly corroded fragments.

The second milestone concerns the carry function for addition and subtraction. It occurred more than 1,700 years later.

Pascal: Reliable Carry Mechanism

Blaise Pascal (1623–1662) was a gifted mathematician, physicist, inventor, writer, and philosopher.[387] Before age 19, he started working on calculating machines to assist his father who, as supervisor of taxes, oversaw a vast number of arithmetical calculations.

The final version of Pascal's calculator, initially called *arithmetic machine* and later *Pascaline*, handled addition via a sophisticated carry mechanism; subtraction via the *9-complement method*, which converts subtraction into equivalent addition;[388] and multiplication and division by repeated application of addition and subtraction. He produced and sold about 20 copies of the machine.[389]

Pascaline signed by Pascal in 1652. Musée des Arts et Métiers, Paris.[390]

The photo of the calculator shows a row of input wheels and, behind it, a row of output slots. Rotating drums placed below these slots produce the displayed output number.

A reset produces 0 for all output slots. When a number is specified by rotation of the input wheels, that number is immediately added

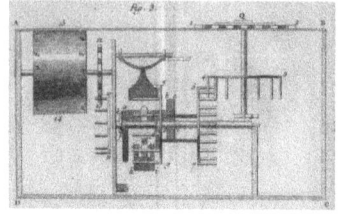

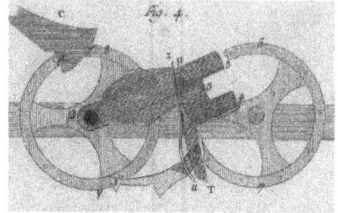

Left: Blaise Pascal, copy of painting by François II Quesnel.[391]
Top: Pascaline input/output for one digit.[392]
Bottom: Pascaline carry mechanism, detail.[393]

to the displayed output value. The carry mechanism is so designed that the digits of a number can be input in any order and do not require unusual force by the operator when multiple carry steps are triggered, such as during addition of 1 to 9,999.

The latter effect is achieved by an ingenious carry mechanism where the input for each digit by the operator effectively stores energy for a potentially occurring subsequent carry step. That energy is stored by gradually lifting a carry gadget when the value for a digit is input.

When the carry step is needed, the gadget drops, and the released energy is used to advance the next digit by one position. Thus, multiple carry steps do not require extra force by the operator.

The Pascaline could be used for multiplication and division by repeated addition and subtraction. But this was a tedious process.

The third milestone is the invention of a calculator that handles multiplication and division directly. It occurred about 40 years after the invention of the Pascaline.

Leibniz: Efficient Multiplication and Division

Around 1672, Leibniz became interested in the construction of a calculator that could carry out addition, subtraction, multiplication, and division with little manual effort.[394]

He was irritated that mathematicians wasted energy on tedious computational tasks, writing:

"[I]t is beneath the dignity of excellent men to waste their time in calculation when any peasant could do the work just as accurately with the aid of a machine."[395]

He realized that any such machine had to be able to store and retrieve multipliers and divisors, a feature absent from the Pascaline.

This could be readily accomplished if the nine cogs of a wheel could be dynamically adjusted downward to eight cogs, or seven cogs, etc., all the way down to zero cogs.

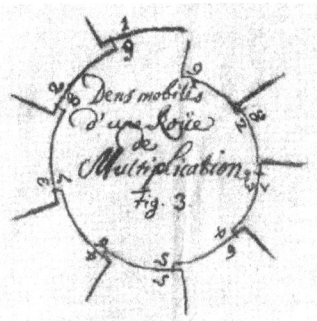

Pinwheel, sketch by Leibniz.[396]

The number of active cogs of such a storage wheel would represent the stored value. It could be retrieved and transferred to another wheel by one full rotation of the storage wheel regardless of the stored value.

In a first attempt at effective multiplication and division, he considered adding a mechanism with storage wheels to the Pascaline.[398]

Model of Leibniz pinwheel.[397]

For this approach, he invented the *pinwheel (Sprossenrad)* where nine pins are stored as spokes in a hub and are selectively pushed out from the center via arc segments to become active cogs. For example, when the arc segment for pin 5 is pushed out, then in a cascade of reactions the arc segments for pins 4, 3, ..., 1 are pushed out as well. Thus, pins 5, 4, ..., 1 become active cogs.

Eventually, he must have thought that the modification of the Pascaline would be too complex since the only surviving evidence of the invention is the above sketch.

Later developments proved the pinwheel to be very effective: It was used in mechanical calculators right up to the advent of electronic calculators in the 1970s.[399]

Leibniz then decided on a new construction that would not rely on the Pascaline. The resulting machine was later named the *stepped reckoner* and in Germany simply *Leibniz'sche Rechenmaschine*.

The only surviving stepped reckoner constructed under Leibniz's supervision. On display at Gottfried Wilhelm Leibniz Bibliothek, Hannover, Germany.[400]

For the wheels with variable number of cogs, he invented another storage wheel mechanism, now called *Leibniz wheel* or *stepped drum* (*Staffelwalze*).

In the picture of the Leibniz wheel included on the next page, the small gear can be moved along its axis. Depending on position, it then engages all nine cogs of the drum gear, or eight cogs, or seven cogs, etc., all the way to zero cogs.

Just as happened for the pinwheel, this second storage wheel withstood the test of time. It was used in mechanical calculators up to the 1970s.[401]

The input for the stepped reckoner is specified by eight dials with ring-shaped knobs.

A telephone-like dial to the right of the input knobs sets the value of one digit of the multiplier. When that value is set to 1, multiplication becomes addition.

Leibniz wheel.[402]

The input module containing the input knobs and the multiplier dial is mounted on rails and is moved laterally via the crank on the left end of the machine. The position of the input module determines the power of 10 applied to the multiplier value.

Ring-shaped input knobs in front, output wheels in center, and carry mechanism in back.[403]

Addition and multiplication are performed by counterclockwise rotation of the crank in front. Subtraction and division are accomplished with the opposite rotation The output is displayed on 16 wheels positioned behind the input module.

Left: Telephone-like dial for multiplier, on top of the input module and to the right of the input knobs.[404]
Right: Carry mechanism, detail.[405]

The various gears in the back of the machine have one cog, five cogs, and ten cogs. The gear with one cog is not visible in the photo. These gears execute the carry steps for the four arithmetic operations without any manual assistance.[406] This contrasts with the Pascaline, whose carry steps accommodate just addition and not subtraction. That restriction may have been one of the reasons why Leibniz abandoned modification of the Pascaline.

At the time of Leibniz, gears could not be produced with sufficient precision for flawless operation of the complicated machine. But a machine built according to modern manufacturing standards at the beginning of the 21st century turned out to compute perfectly, including the operation of the sophisticated carry mechanism.[407] In particular, no manual interference has ever been required in extensive tests of the four arithmetic operations.[408]

Leibniz is the inventor of the binary system. Intrigued by the simplicity of the four arithmetic operations for these numbers, he designed, but did not implement, an unusual calculator for binary addition and multiplication that relies on balls, holes, and channels to control the flow of the balls.

The machine has an array with holes that can be opened and closed. Each hole corresponds to one digit of a binary number. The holes are opened at the positions corresponding to a 1 and are closed corresponding to a 0. Small balls fall through the open holes into channels and interact with other balls in such a way that binary multiplication is carried out.

In addition, Leibniz invented a machine for conversion of decimal numbers to binary ones.[409]

It is unfortunate that Leibniz didn't pursue binary computation beyond the description of machines. Had he done so, he would have fully appreciated the extraordinary reduction in machine complexity.

Binary calculator based on Leibniz's description. Arithmeum, Bonn.[410]

Indeed, the results would have triggered the age of binary machine computation. Instead, that change came 250 years later.

We advance 150 years to the fourth milestone, where a conceptual machine executes sequences of the four basic arithmetic operations according to specified rules. The machine was completely designed but never implemented. It constitutes the first computer.

Babbage: First Computer

Charles Babbage (1791–1871) was an engineer, inventor, mathematician, and philosopher with wide-ranging interests.[411]

For example, he invented the *dynamometer car* in 1838 to record the performance of a train, such as speed, pulling force by the engine, and shaking of carriages.[412] As the train proceeds, the data are recorded by a pen on a continuously moving roll of paper.

He came to invent the dynamometer car with its recording equipment so that the controversy raging at the time about the most desirable width of railroad tracks could be scientifically settled. Thus, he replaced intuitive reasoning about stability and shaking of carriages with a decision based on objective data.

The dynamometer car spawned numerous recording devices for transportation equipment, such as the flight recorder[414] now employed in aviation.

The most important inventions of Babbage concern computation. Up to Babbage's time, computing equipment re-

Charles Babbage.[413]

quired manual input for each arithmetical step. He introduced a new view: Computing equipment shouldn't just perform individual computational steps, but should be able to carry out complicated sequences.

In a first move toward this goal, Babbage invented the *difference engine* in 1822. It carried out complicated arithmetical operations such as computation of logarithms.[415] The machine printed tabular output to eliminate human, and hence error-prone, transcription of results.

The difference engine relied on a large number of gears stacked on vertical shafts and required extraordinary craftsmanship and precision during construction.

The British government agreed to fund the project and in total supplied £17,000. It is difficult to estimate the equivalent cost today, but a reasonable estimate for 2016 might be five million US dollars.[416]

Babbage could not complete the implementation of the difference engine due to several factors, not the least of which were the technical complexity of the machine and the comparatively inefficient tools to manufacture the large number of components. It is estimated that it would have been composed of around 25,000 parts weighing 15 tons in total.[417]

In the period 1847-49, he redesigned the machine and produced *difference engine no. 2*, which was substantially simpler. It also wasn't constructed during Babbage's lifetime, but a determined effort in the late 20th and early 21st century created two working engines.

Difference engine no. 2. Science Museum, London.[418]

The construction used materials and engineering tolerances of the 19th century, thus proving that a complete engine could have been built during Babbage's time. Each completed machine has about 8,000 parts and weighs five tons, a significant reduction from the complexity and weight of the original difference engine.[419]

While working on the design of the difference engine, Babbage developed a vision of a much more complex and powerful computing machine. He called it the *analytical engine*.

It had all the features of modern computers: storage of numbers; a unit for computations; a section for control of operations; and an input/output system using punched cards based on Jacquard's automated loom.[420]

He hired a first-rate draftsman with considerable engineering knowledge, C. G. Jarvis, to create about 300 engineering drawings. They were accompanied by hundreds of large sheets Babbage called *mechanical notations* that explained function and performance.[421]

Babbage built a trial model of part of the engine, but due to its complexity and huge construction cost, the entire machine was never built.[422]

The Analytical Engine

Trial model of a part of the analytical engine, built by Babbage. Science Museum, London.[423]

Babbage viewed the analytical engine as a powerful tool for complicated numerical computations.

A broader view was developed by mathematician and writer Ada Lovelace (1815–1852), who met Babbage as a teenager in 1833 and became fascinated with the analytical engine.[424]

Over many years, Babbage and Lovelace carried out an extensive correspondence about the analytical engine
and its role and impact.

In 1842-43, Lovelace translated notes
written by Luigi Federico Menabrea
(1809–1896) about the analytical engine from French to English.

Menabrea had compiled them after
Babbage had given talks about the
engine at the university of Turin in
1840.

Ada Lovelace, by Margaret Sarah
Carpenter, detail.[425]

Lovelace felt that Menabrea's expo-
sition should be amplified by additional explanations. To that end,
she appended a section called *Notes by the Translator* to the translated article. The modest title notwithstanding, the added section
is about three times as long as the translated text of Menabrea's
article.

In subsections called *Note A* through *Note G*, she develops fundamental concepts of what is now called computer programming.
In particular, in a clear and concise exposition she lays out how
specification of operations and variables of such a program is very
different from the standard use of operators and variables in mathematics.

She also envisions a far greater role for the analytical engine than
just the computation of numbers. In her words,

"[The analytical engine] might act upon other things besides *number*, were objects found whose mutual fundamental relations could
be expressed by those of the abstract science of operations, and
which should be also susceptible of adaptations to the action of
the operating notation and mechanism of the engine.

"Supposing, for instance, that the fundamental relations of pitched
sounds in the science of harmony and of musical composition were

susceptible of such expression and adaptations, the engine might compose elaborate and scientific pieces of music of any degree of complexity or extent."[426]

The prediction of music composition by the analytical engine must have seemed far-fetched at the time of Lovelace. But today computers are used in many ways to create or process music.[427]

Thus, the modestly named *Notes by the Translator* make Lovelace not only the first-ever computer programmer, but also a visionary of future uses of computers.

We pause the discussion of practical computing machines and turn to a theoretical development about computing that constitutes the fifth milestone. It occurred in the mid-1930s.

Turing: What can be computed?

For centuries, it was assumed implicitly that the values of any function, no matter how contrived, could be computed.

In 1935, Turing set out to challenge this notion. To this end, he needed a precise definition of computation that would apply not only to past machines, but to *all* machines that could ever be constructed.

Of course, that goal was unattainable: Who could anticipate all machines made in the future? So instead Turing focused on a very simple machine, now called the *Turing machine*, that was easy to understand, yet by suitable steps could simulate any complicated computing machinery he could dream up.

The machine consists of a tape of infinite length, a head that reads entries on the tape, and a table of instructions. The tape is subdivided into cells.

Initially, the tape is blank, optionally denoted by a 0 in each cell, and some instruction of the table is defined to be the current one. Thus, the head is positioned over a blank cell.

Then an iterative scheme proceeds as follows. The head reads the value of the cell over which it is positioned. Depending on the acquired value and the current instruction, the following happens: The head writes a 0 or 1 into the cell, shifts the tape to the right or left by one cell, and finally defines some instruction of the table to become the current one.

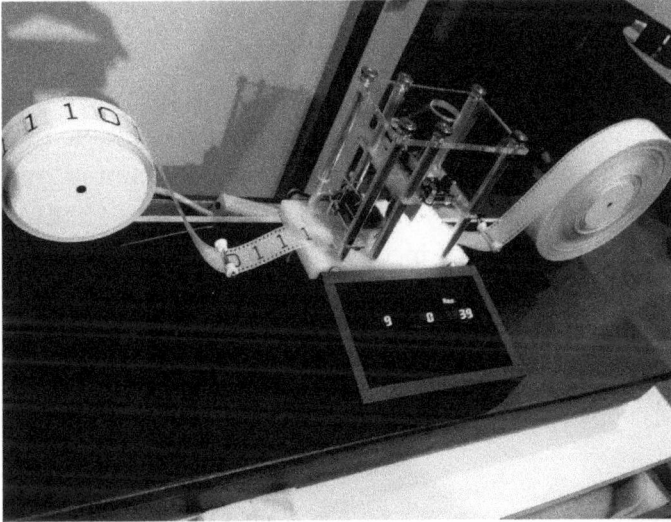

Implementation of the Turing machine. The infinite tape is approximated by a long tape stored on two spools[428]

This trivial machine is very inefficient, but—amazingly enough— can compute whatever the most powerful computers accomplish today. Obviously, those computers also can do what the trivial Turing machine accomplishes, so in some sense they all have the same computing power.

Due to this fact, the Turing machine has become the standard for computing capability: If computational equipment can do what the Turing machine can accomplish, then that equipment is declared to be *Turing-complete*. Except for Babbage's analytical engine, none of the machines discussed so far are Turing-complete.

Recall that in the 1930s Gödel established two incompleteness theorems that forever limited what mathematics could accomplish.[429]

To these results, Turing added limits on computing using his ingeniously simple machine.[430] Indeed, that machine and variations defined later became powerful tools for proving a variety of results, and to this day are central devices for investigation of computational power.

At the time these limits were established for mathematics, computing technology took a giant step forward. The key person in this development was Konrad Zuse (1910–1995).

He realized that computation with binary numbers on a mechanical computer was eminently feasible and far preferable to decimal computation. Indeed, he correctly viewed the required translation of decimal input to binary numbers and of binary results to decimal output to be a minor inconvenience when compared with the extraordinary simplification brought about by binary data storage and binary arithmetic. This leap from decimal to binary computation constitutes the sixth milestone.

Zuse: First Functioning Computer

In 1935, just one year after obtaining a degree in civil engineering, Zuse quit a promising job at Henschel Flugzeugwerke, a major aircraft manufacturer. He was bored by the extensive manual computations of his job and decided to design and build a fully automated calculating machine.[431]

Zuse was aware of Leibniz's result that binary numbers could be used for computations instead of decimal

Konrad Zuse.[432]

numbers, and he knew propositional logic. But he didn't know about Babbage's analytical engine, Lovelace's development of computer programs, or Turing's results about computation.

His key decision was that all computations involving binary numbers were to be based on principles of propositional logic.

In 1936, he began construction of the first computer in the living room of his parents' apartment in Berlin. The computer, later called Z1, was completely mechanical: conversion from external decimal to internal binary numbers and vice versa, storage of numbers, program control by punched tape, and binary computations.

Z1 computer, constructed in parents' living room, Berlin, 1938.[433]

In particular, he developed a binary encoding of real numbers as *floating point numbers*[434] and a module for processing them. The encoding has become the standard method for any computing device processing real numbers.

The main construction material was sheet metal. About 30,000 sheet-metal pieces were produced in a painstaking manual effort. An electric motor taken from a vacuum cleaner powered the computer.

Several friends helped with the fabrication of the numerous components. Initial financial support was provided by his family and friends. But by 1937, additional financing was needed, but difficult to obtain.

When Kurt Pannke, the former president of a company producing specialty calculators, was asked for financial support, his first reaction was, "I have been told that you have invented a calculator. I don't want to discourage you, to work as inventor and develop new ideas. But I must say the following up front: For calculators, all possibilities have already been explored. Nothing new can be invented any more."

But after a visit with Zuse, Pannke became convinced that something very different had been invented, and provided substantial financial support.[435] By 1938, the Z1 had been completed.

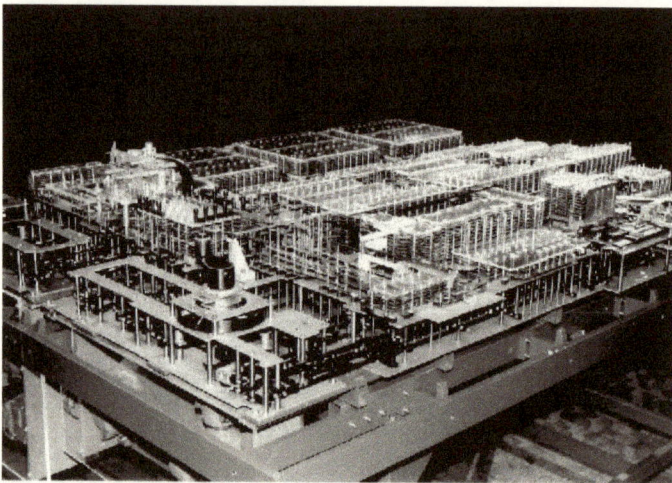

Reconstruction of Z1 computer, 1989. The original Z1 was destroyed in World War II.[436]

It was an amazing accomplishment: Here was a computer with modules for input, output, storage, control, and computing, made from the simplest possible material. And development and construction were completed in less than three years!

A recurring problem in the operation of the Z1 was jamming of the sheet-metal pieces. So already during the construction of the Z1, Zuse began work on a different approach.

A key collaborator of this effort was Helmut Schreyer (1912–1984), who had a strong background in electromechanical and electronic equipment such as relays and vacuum tubes. The outcome was computer Z2.

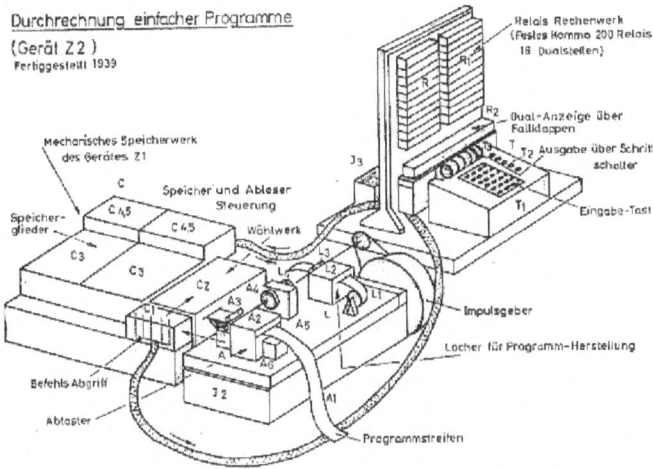

Drawing of Z2 computer. The Z2 was destroyed in World War II.[437]

It carried out computations with electromagnetic relays, but for data storage used the mechanical module of the Z1.

The Z2 was a proof-of-concept machine. Its successful operation convinced Zuse to start construction of a new machine, called Z3, that used relays throughout for storage and computation.

Reconstruction of Z3 computer, ca. 1961. The original Z3 was destroyed in World War II.[438]

Completed in 1941, the Z3 was the world's first programmable, fully automatic digital computer.[439] The program instructions were punched into 35mm film.

Thus, in less than six years, Zuse had started with knowledge about binary computation and propositional logic, and had produced the world's first fully functioning computer.

The effort did not stop with the Z3. A much more powerful machine was planned, the Z4.[440] Zuse began construction in 1942. Despite heavy bombing of Berlin, he managed to save it past the end of World War II, and completed it by 1949.

Original Z4 computer. Deutsches Museum, Munich.[441]

It was purchased in 1950 by the Eidgenössische Technische Hochschule (Swiss Federal Institute of Technology) in Zurich, making it the only commercially used[442] programmable computing machine in Europe. The Z4 ran in Zurich for five years. In 1954, it was transferred to the Institut Franco-Allemand des Recherches de St. Louis

(Franco-German Institute of Research St. Louis) in France, where it was in use until 1959. Today, the Z4 is on display at Deutsches Museum in Munich.

While the Z4 miraculously escaped destruction during the bombing of Berlin in World War II, all of its predecessors were destroyed. With Zuse's assistance, much was later reconstructed, including the machines Z1 and Z3.

His son Horst Zuse (1945–) has assembled an extensive library covering Zuse's life, his pioneering achievements, and the extensive reconstruction effort.[443]

If we were to treat the entire history of computing in this chapter, the next milestone would be the invention of the electronic computer. As explained earlier, we won't proceed to that event or subsequent milestones since any discussion within the confines of one chapter would be superficial at best.[444]

Summary

The chapter covers several stages in the development and implementation of computing machines. We see that the human mind not only is able to develop mathematics, but also can design equipment carrying out complex mathematical computations.

The design work was done under severe restrictions imposed by limits of available materials and crude production methods. Thus, it is one more demonstration of extraordinary human creativity and innovation.

We have reached the end of the first part of the book. The second part investigates a basic philosophical question about mathematics: Is it created or discovered? It may seem that the book thus consists of two distinct and rather disjoint parts. But this is not the case: The history of mathematics turns out to be essential for certain arguments made in the second part.

8

Question: Creation or Discovery?

In Chapters 2–6, we have seen how various ideas and results of mathematics were developed—not in a straight line of ever better insight, but with detours and false turns. Eventually, clarity prevailed, giving us today a body of mathematics of mind-boggling complexity, beauty, and utility.

The question arises: Where do all these wonderful results of mathematics come from? Put differently: Were they created by the human mind, or were they always present in some realm and then discovered? In brief: Is mathematics created or discovered?

The answers to that question vary[445] greatly, including the claim that both creation and discovery take place. Shortly we will see the clashing viewpoints of several eminent mathematicians.

We need a precise definition of the words "create" and "discover" to avoid confusion later. We always use the word "create" to mean "bring something into existence," and the word "discover" to denote "detect an existing thing."

For example, the amazingly simple yet very effective screw pump for lifting water was *created* by Archimedes,[446] and the Constitution of the United States by the Constitutional Convention.[447] On the other hand, the continent Antarctica was *discovered* by several expeditions[448] in 1820, and the double helix of DNA by James Watson and Francis Crick in 1953.[449]

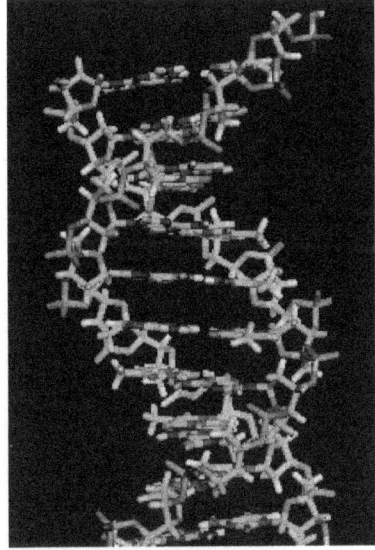

Left: Modern Archimedes screws drain low-lying land in the Netherlands.[450]
Right: Double helix DNA.[451]

The above definitions agree with those of various dictionaries.[452] But they also rule out certain use that would cause confusion.

For example, in the statement "The impressionist painters discovered a new way to paint shadows of a sunlit scene using a bluish tint"[453]—see the next page for an example—the claimed discovery does not involve an existing thing, but something new. In fact, we could replace the word "discovered" by "created" without changing the meaning. We will avoid such confusion by using "discover" only if the discovered thing exists already.[454]

Despite the widely varying claims about creation or discovery of mathematics, or for that matter, about acceptance or rejection of axioms and proof processes, there is almost universal consensus among mathematicians on one point: Counting and the natural numbers are the start of arithmetic, indeed of all of mathematics. This agrees with the historical record, as sketched in Chapter 2.

For example, statements by Dedekind and Weyl below express that belief, even though Dedekind was a proponent of the axiomatic

Impression, Sunrise, by Claude Monet, 1872, detail. The painting became the source of the label *impressionism*. For a large color reproduction, see Wikipedia "Claude Monet."[455]

method of mathematics, and Weyl was an intuitionist at the time he made the statement.

Dedekind: "I view all of arithmetic to be a necessary or at least natural result of the most elementary act, the counting. And counting is nothing but the successive creation of an infinite sequence of the positive whole numbers in which each individual is defined from its immediate predecessor."[456]

Weyl: "Mathematics starts with the sequence of natural numbers, that is, the law that from nothing creates the number 1 and from each already created number the successor."[457]

There are two subtle points.

First, Dedekind does not specify who does the counting, and Weyl does not say where the cited law comes from.[458] Chapter 10 provides a different explanation for the origin of mathematics that does not rely on a priori assumptions.

Second, when we count things such as apples, we implicitly declare the differences between any two apples—for example, of color or size—to be irrelevant and thus consider all apples to be the same.

In mathematics, this idea is made precise with the definition of *equivalence class*,[459] where any two items of the class are declared to have the same characteristics; or rather, where any distinguishing characteristics are ignored.

It is safe to say that mankind got the intuitive idea underlying the mathematical concept of equivalence class by evolution. For example, people who didn't recognize an approaching lion as dangerous if they had not seen that particular animal before, were simply eaten.

The review of prior answers begins with Plato (424(?)–348(?) BCE). He is the pivotal figure in the development of Western philosophy. As Whitehead put it,

"The safest general characterisation of the European philosophical tradition is that it consists of a series of footnotes to Plato."[460]

Plato. Luni marble, copy of the portrait made by Silanion ca. 370 BCE for the Academia in Athens. From the sacred area in Largo Argentina.[461]

Plato: Realm of Abstract Objects

We focus on one postulate of Plato, called *Platonism*, that is relevant for the question of creation versus discovery of mathematics.

It claims that abstract objects exist in a realm distinct from the external world and from the internal world of consciousness.[462] For example, Plato writes,

"Geometers, arithmeticians, and astronomers are in a sense seekers since they do not produce their figures and other symbols at will, but only explore what is already there."[463]

A modern version of Platonism that specifically applies to mathematics is called *mathematical platonism*. Note the lower case "p," which differentiates Plato's original formulation from the modern version. Mathematical platonism says that *all* of mathematics resides in a realm of abstract objects that is separate from the sensible world. Mathematicians then discover some of these objects and state them as axioms, theorems, and so on.[464]

Mathematical platonism represents the strongest vote for discovery, in the sense that all of mathematics is considered to be in a realm of abstract objects.

One could investigate weaker versions, for example, by postulating just the natural numbers to be in a realm of abstract objects and declaring all mathematical results built upon them to have been created.[465]

Plato's concepts dominated philosophy for more than 2,200 years. As a result, mathematics was thought to be discovered, possibly with divine help. But from the 19th century onward, some eminent mathematicians voiced the opinion that mathematics is created by the human mind, while others defended discovery. The next section contains a sample of these opinions.

A Sample of Opinions

We begin with Gauss, who evidently believes in creation of mathematics: "[N]umber is purely a product of our mind."[466]

Cantor shares that view: "The essence of mathematics lies entirely in its freedom."[467]

Dedekind firmly states that the negative and rational numbers were created: "[T]he negative and rational numbers are created by man."[468] He declares the irrational numbers to have been created as well when he titles a section describing construction of the irrational numbers with "Creation of the irrational numbers."[469].

A slightly different evaluation is made by Kronecker. He believes that the integers are given a priori by a metaphysical construction, but that the rest of mathematics is created by man: "God made the integers; all else is the work of man."[470] He needs the metaphysical construction of the infinite number of integers: As a believer in finitism, he cannot delegate that task to man.

The opposite view is offered by Frege: "I hope I may claim in the present work to have made it probable that the laws of arithmetic are analytic judgments and consequently a priori."[471]

Gödel also is in favor of discovery: "I am under the impression that after sufficient clarification of the concepts in question it will be possible to conduct these discussions with mathematical rigour and that the result will then be ... that the Platonistic view is the only one tenable."[472]

Frege and Gödel do not claim their conclusions with certainty. Frege says "I hope I may claim ...," and Gödel states "I am under the impression that ...." They consider their conclusions inescapable, yet they are unable to firmly establish them the way mathematical theorems are proved.

The controversy continues to this day. Below are two diametrically opposed statements by Roger Penrose (1931–) and Armand Borel (1923–2003), made at the beginning of the 21st century.

Penrose argues for discovery using Plato's original concept of a realm of abstract objects: "Objective mathematical ideas must be thought of as time-

Roger Penrose.[473]

less entities and are not to be regarded as being conjured into existence at the moment that they are first humanly perceived."[474]

Indeed, he describes detailed relationships between the physical world, the mental world—which is the world of perceptions—, and the Platonic mathematical world.[475]

In contrast, Borel declares mathematics to be simultaneously an art and a science created by the human mind: "[Mathematics] is an art because it is primarily a creation of the mind, and progress is achieved by intellectual means, many of which issue from the depths of the human mind and for which aesthetic criteria are the final arbiters. ... [Mathematics also] is a *mental* natural science ... whose objects and modes of investigations are all creations of the mind." [emphasis in the original][477]

Armand Borel.[476]

In the next chapter, we see how one can investigate philosophical conundrums such as the question about creation or discovery of mathematics, and eventually obtain clarity. The process does not involve constructing a proof as alluded to by Gödel, but viewing the problem from many angles until enough insight is gained to resolve the problem.

Summary

There has been a multitude of claims for creation or discovery, as well as for concepts in between.

On the surface, the arguments are made with mathematical precision.[478] Yet, key conclusions drawn by various persons are in conflict—a strange fact. Later in this book, we see how this is possible.

Let's avoid discussions involving various abstract terms and investigate the question "Creation or discovery?" outside philosophy, if the reader can imagine such a thing.

Of course, we need lots of help with that approach. We will get it in the next chapter from a person who has been declared a philosopher, but who always said that philosophy is not a profession: Ludwig Wittgenstein.

We use that insight in subsequent chapters, where we explore the question of creation or discovery. The ultimate conclusion of those considerations is: *All of mathematics is created and none is discovered.*

9

Wittgenstein's Philosophy

Ludwig Wittgenstein (1889–1951) is often considered to be the most eminent philosopher of the 20th century.[479] He developed entirely new ways of looking at philosophical questions and used them to achieve impressive results.

In particular, he showed that the human language has limits that we almost always violate when investigating philosophical questions.

When we commit such errors, we produce sentences that, though perfectly grammatical and seemingly coherent, are actually nonsensical. One might say that the human language contains traps we almost surely fall into when dealing with philosophical questions. These traps are so insidious that we

Ludwig Wittgenstein.[480]

typically do not find a way out. Wittgenstein proposes the image of a fly trapped in a fly bottle[481] to characterize the situation.[482] Trying to find an exit, the fly keeps on circling, but never finds the way out. Wittgenstein advises how we can escape such traps. A key tool invented by him for such escapes is the language game.

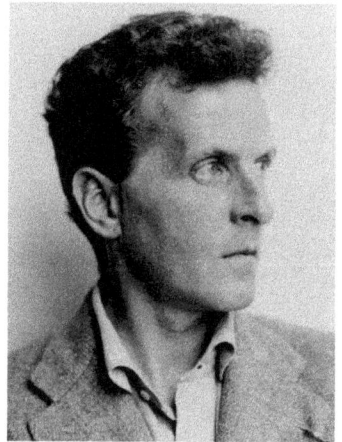

Language Game

We investigate some questions and their answers to demonstrate the use of Wittgenstein's language games.

First, somebody asks "Is Rome the capital of Italy?" The obvious answer is "yes."

Second, "Does Rome lie within a 10-mile radius of Berlin?" The answer is "no," as is readily proved with a map of Europe.

Third, "Is Rome east of voltage?" This is an odd question. Though the syntax of the question is correct, the cited city, direction, and concept of electricity are assembled in a way that does not make sense. So we do not answer "yes" or "no," but declare that the question is *nonsensical*. In contrast, the first two questions have a meaningful interpretation and thus are *valid*.[483]

The analogous classification applies to statements. That is, a statement is *nonsensical* if its terms collectively do not make sense, and is *valid* if it has a meaningful interpretation.[484]

A valid statement may be false; an example is "The moon is made of cheese." In contrast, a nonsensical statement such as "Sun plus moon equals time" cannot be evaluated at all.

Let's consider a more difficult question: "Is black a color?"[485]

To investigate the possible answers, our brain may bring up a recent situation where we were in the basement, the light bulb went out, and suddenly we were engulfed in blackness. Since that effect was produced by absence of light and colors, we may claim, "No, black is not a color."

Another person may recall the recent purchase of a car where black was selected as the desired color of the car. So that person says, "Yes, black is indeed a color." At this point, a lively discussion ensues, with logic arguments flying back and forth.

This is an example situation where a philosophical question results in conflicting answers. Yet, each of the answers is based on seemingly unassailable logic arguments. A subsequent discussion may go on and on in a cycle of repeated arguments. When the participants eventually stop due to fatigue, nothing has been resolved. It is an instance of the fly circling in the fly bottle.

How can this seemingly irreconcilable conflict be resolved?

For insight into the situation, let's review how the human brain comes up with answers to philosophical questions.

The brain first establishes the meaning of the components of the question and their implied relationships. This is done seemingly instantaneously and without conscious awareness. Once that interpretation is at hand, the brain carries out conscious reasoning toward an answer.[486]

The steps include an implicit check for consistent use of concepts and terms in the question. That check is easy for the question "Is Rome east of voltage?" The brain correctly declares the question to be nonsensical and stops all reasoning about it.

But when the brain faces philosophical questions, detection of nonsensical cases is much more difficult. Thus, the brain may carry out logic reasoning that cannot possibly produce a relevant result. The question "Is black a color?" is of that variety.

The failure of the brain to unmask a philosophical question as nonsensical may be due to a number of factors. Leading among them is consideration of an insufficient number of situations in which the question may arise. As a cure for this defect of the brain, Wittgenstein invented the method of *language games*.[487]

Each such game is defined by some language use and action imagined to take place in the world. Thus, the game is not just some abstract construct.

Furthermore, each game is so selected that in some sense it is connected with one facet of a given philosophical problem.

Wittgenstein *operates* a language game by virtually living in the environment of the game and observing its operation. Since the language game takes place in a specific setting of the world, it is guaranteed that any insight so gained is well-grounded.

We could say that the operation of the language game produces *experiences about the world* in the brain that are relevant for understanding one facet of the question.

The notion of *experiences about the world* in the brain is made precise in Chapter 13, where we use modern brain science to understand the brain's processing of philosophical questions and statements.

When the creation and operation of language games is carried out for all important facets of the given philosophical problem, the brain is empowered by the numerous experiences and thus is able to resolve the philosophical question.

Here, "resolve" does not mean that an explicit solution or explanation based on logic arguments has been found that proves the problem to be nonsensical. Instead, "resolve" indicates that the philosophical problem is no longer baffling and has become clear, and thus does not require further investigation. We could say that this insight shows the question to be nonsensical. In an alternate reaction, we may choose to simply ignore the question from now on.

For a demonstration of the process, consider the following philosophical statement:

"The individual words in language name objects, sentences are combinations of such names .... Every word has a meaning. This meaning is correlated with the word. It is the object for which the word stands."[488]

For investigation of these claims, Wittgenstein defines the now-classic *builder's game*:[489]

"The language is meant to serve for communication between a builder *A* and an assistant *B*. *A* is building with building-stones: There are blocks, pillars, slabs, and beams. *B* has to pass the stones in the order in which *A* needs them. For this purpose they use a language consisting of the words 'block,' 'pillar,' 'slab,' and 'beam.'

A calls them out;—*B* brings the stone which he has learnt to bring at such-and-such a call.—Conceive this as a complete primitive language."

Wittgenstein explores a child's acquisition of language using this language game. To this end, he imagines that the specified commands constitute the *entire* language of *A* and *B*, indeed of a tribe. The children learn to carry out *these* activities, use *these* words, and react in *this* way when others say them.

Wittgenstein's investigation of this game then produces insight into the philosophical problem of language acquisition.[490] In the process he resolves the above-cited philosophical claims; indeed, it becomes evident that the claims are untenable.

Wittgenstein used his technique to resolve a number of complex philosophical claims and problems.[492] We will look at one of them later in this chapter: the theory of perception of color proposed by the famous German poet Johann Wolfgang Goethe (1749–1832), who believed it to be a crowning intellectual achievement. Yet, Wittgenstein proved the theory to be untenable. In fact, he showed that there cannot be any consistent theory of the type proposed by Goethe.

Johann Wolfgang Goethe, by Joseph Karl Stieler, 1828, detail.[491]

Wittgenstein not only proposed language games as powerful tools for the resolution of philosophical problems, but also employed them in such a lucid way that their discussion contains veritable templates for other philosophical investigations.

We use these templates when we examine the question of creation versus discovery of mathematics in subsequent chapters.

Wittgenstein didn't begin his investigation into philosophy with the concept of language games. He first created a very different approach rooted in logic. A number of years later, he recognized that the proposed theory contained "grave mistakes."[493] He then started a very different investigation that led to the concept of language games. The next section sketches that journey.

The Journey

Wittgenstein's early efforts in philosophy, carried out in the 1910s, resulted in the famous book *Tractatus Logico-Philosophicus*.[494] The title typically is abbreviated to *Tractatus*.

The book provided an all-encompassing analysis of language and meaning. Indeed, Wittgenstein claimed that the results forever banned contradictory and confusing philosophical arguments by proving them to be nonsensical.

The arguments of the Tractatus are too complicated to be included in this chapter; the Notes include a summary.[495] But there is no need for us to get into the material of the Tractatus anyway, since we will rely on Wittgenstein's later work, begun in the second half of the 1920s.

At that time, he realized that basic premises of the Tractatus were untenable[496] and embarked on a new effort that continued until his death in 1951. He summarized his thoughts in a second book, the *Philosophical Investigations*.[497] It introduced and extensively used language games. The book was almost completed at the time of his death in 1951 and was published shortly thereafter.

Wittgenstein also left a vast number of notes. They not only help us understand the Philosophical Investigations, but also are important contributions to philosophy in their own right. That material was published posthumously in several books.[498]

The above discussion about language games includes the claim that they help *resolve* philosophical problems. We now look at resolution in more detail.

Resolution of Philosophical Problems

The core claim of the Tractatus is that valid statements of language must be connected with the world in a logically consistent manner.[499] Any other statement is nonsensical, which in Wittgenstein's view means that such statements should not be uttered.[500]

The Tractatus validates this claim using arguments of mathematical precision. But the statements of the Tractatus themselves are not connected with the world in the required fashion and thus, by the very arguments of the Tractatus, are nonsensical!

Wittgenstein comes to this conclusion toward the end of the Tractatus. In paragraph 6.54 of the Tractatus, he suggests the following remedy:

"My propositions are elucidations in this way: Anyone who understands me eventually recognizes them as nonsensical, when he has used them—as steps—to climb up beyond them. (He must, so to speak, throw away the ladder after he has climbed up on it.)

"He must transcend these propositions, and then he will see the world the right way."[501]

When Wittgenstein abandoned the Tractatus, he also lost the justification for declaring most philosophical state-

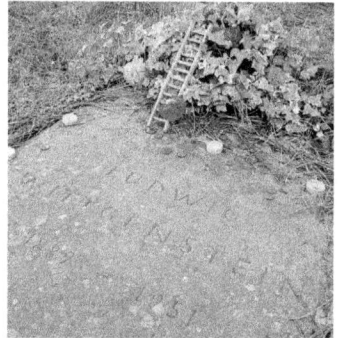

Wittgenstein's gravestone. The ladder is part of small mementos left at the grave by visitors. It refers to statement 6.54 of the Tractatus.[502]

ments to be nonsensical. But he still believed that nonsensical statements of philosophy should never be uttered.

So when he used language games to resolve a particular philosophical problem, he typically described just each language game and its operation, but didn't compare the insight with nonsensical claims made by others in connection with that problem. Indeed, any comparison statement would have been nonsensical as well!

Thus, it was up to the reader to draw the desired conclusion from the information supplied by the language games.

By this approach, Wittgenstein did not utter nonsensical statements, as desired. But there was a price to be paid: Many if not most people were not willing to follow that line of reasoning; as a result, his work was controversial.

There is a way out of this dilemma that retains the full power of language games, yet does not leave it to the reader to guess what is to be concluded from the operation of a particular language game: The language game is listed together with a philosophical statement or claim. Conclusions from the language game then establish that the statement is nonsensical or, simpler, that the statement should be ignored.

For example, we might investigate the philosophical statement "The world is not real; only our perceptions are real."[503] A part of that assertion is the claim "The world is not real." We then describe language games that address that claim and ultimately show it to be nonsensical. We deal similarly with the claim "Only our perceptions are real."

One may declare that this method violates Wittgenstein's principles since it explicitly lists nonsensical claims. Well, not quite. In the introduction to the Philosophical Investigations,[504] he writes,

"Two years ago I had occasion to re-read my first book (the *Tractatus Logico-Philosophicus*) and to explain its ideas to someone. It suddenly seemed to me that I should publish those old thoughts and the new ones together: that the latter could be seen in the right light only by contrast with and against the background of my old way of thinking."

Instead of printing the two books together, one could list claims of the Tractatus as titles of language games. The insight gained from the language games then would show the claims to be nonsensical.

The next section demonstrates the process using Goethe's *Zur Far-benlehre*[505] (Theory of Color) and language games of Wittgenstein's

Remarks on Colour.[506] That is, the former book supplies the claims, and language games taken directly from or inspired by the latter book refute them.

Goethe's Theory of Color

Goethe's theory of color covers human interpretation of and reaction to light and color with a precision reminiscent of mathematics.

For example, he sharply distinguishes between white, black, grey tones, and colors such as red, green, and blue.[507]

In contrast, Wittgenstein says the following:[508]

"I see in a photograph (not a colour photograph) a man with dark hair and a boy with slicked-back blond hair standing in front of a lathe, which is made in part of castings painted black, and in part of smooth axles, gears, etc., and next to it a grating made of light galvanized wire.

"I see the finished iron surfaces as iron-coloured, the boy's hair as blond, the grating as zinc-coloured, despite the fact that everything is depicted in lighter and darker tones of the photographic paper."

Thus, we may see color when a picture contains none.

The converse occurs, too: We may convert color to black or white.

For example, on a chess board with light brown and dark reddish-brown squares, we see several rosewood chess pieces.

Chess pieces.[509]

Prior to a game of chess, say when comparing the styles and materials of different chess sets, we may declare the pieces to be rosewood colored.

But during the game of chess, we call them "black" and not "dark brown" or "rosewood-colored."

Let's turn to the perception and interpretation of colors in paintings. Suppose we participate in a guided tour called "Impressionist Paintings." We stand in front of Claude Monet's famous painting *Woman with a Parasol*; Wikipedia"Claude Monet" has a large color reproduction. The tour guide may say "The entire pastoral scene is suffused with sunlight. Everything glows. Even the yellow jacket and the white skirt in the shadow of the umbrella are vivid."

Woman with a Parasol (Madame Monet and her son), by Claude Monet, 1874. Patches with bluish tint help produce impression of shadows[510].

Consider another guided tour, called "Impressionism Explained." Standing in front of the same painting, the guide may include in the analysis, "Notice how the jacket and the skirt are painted. There are light blue patches within the yellow of the jacket and the white of the skirt. These patches produce the impression of shadow for the jacket and the skirt much better than a darkened yellow and grey ever would."

On the first tour, the light blue patches of the dress are seen to be yellow for the jacket and white for the skirt. On the second tour, they are declared to be light blue.[511]

Goethe's theory of color has only one interpretation for a color patch in a given setting, so it cannot accommodate the description of color of one of the tours.

A painter sometimes engages in both interpretations in rapid sequence. Mixing paints and brushing them on the canvas, he sees their colors as labeled on the tubes of paint. So light blue paint seen on the brush has come from two tubes, say labeled white and azure. When he has applied paint to the canvas, he steps back and views the effect of the colors on his interpretation. At that time, a patch of light blue may be seen as yellow or white, as is the case for Monet's painting.

The painter Paul Cézanne referred to that process when he exclaimed, "But what an eye Monet has, the most prodigious eye since painting began! I raise my hat to him. As for Courbet, he already had the image in his eye, ready-made. Monet used to visit him [Courbet], you know, in his early days .... But a touch of green, believe me, is enough to give us a landscape, just as a flesh tone will translate a face for us."[512]

Summary

We have explored Wittgenstein's work in philosophy: his early work, the Tractatus; his reason for rejecting it later; and finally, his resolution of philosophical problems using language games as described in the Philosophical Investigations and a number of posthumously published books.

As an example, we have investigated Goethe's theory of color using language games of photography, the game of chess, and impressionism.

The next chapter begins the investigation into the question of creation or discovery of mathematics. The main idea of the chapter is to reinterpret parts of the history of mathematics by reformulating them into language games.

We rely directly and indirectly on ideas of Wittgenstein's book on the foundation of mathematics.[513] Originally, the book generated a rather negative reaction.[514] But an evaluation of the 2010s declares, "Ludwig Wittgenstein's Philosophy of Mathematics is undoubtedly the most unknown and underappreciated part of his philosophical opus."[515]

10

Language Games of History

Chapters 2–6 describe the history of several key concepts of mathematics. In this chapter, we look at some of that material again, but this time in the form of discussions and dialogues that tell how or why mathematicians came upon those concepts.[516]

We sometimes rely on information recorded at the time, but mostly on reasonable guesses on our part. In the terminology of Wittgenstein, we mostly create language games about the human side of mathematical developments. In the process, we gain insight into the philosophical problem of creation versus discovery of mathematics. We begin by reexamining the start of mathematics.

Origins of Mathematics

In the distant past, pebbles or scratches or other marks were used to record a quantity.[517] This could be done without any number concept; each pebble or mark would simply stand for one item.

For example, pebbles would represent the sheep of a herd leaving the village in the morning. When the herd returned from the pasture in the evening, the pebbles were matched with the sheep. When all pebbles had been used up, the entire herd had been accounted for.

To simplify the process, names were made up for groups of pebbles or scratch marks. This led to counting and the natural numbers.

Next, commercial trades and exchanges demanded that collections of items be combined or split up. Rules for handling theses steps became addition and subtraction of natural numbers, and later multiplication and division.

Suppose we had asked the early contributors of this process, "Do you get these ideas on your own, or does your brain somehow look them up?" They would have looked at us with disbelief that we would ask this. "Of course," they would have said, "we worked out how to handle these names and rules, just as we worked out how to handle fire, build huts, or forage for food."

Straightforward steps derived the integers and the rational numbers from the natural numbers. With these concepts at hand, let's look at the definition of the real numbers.

Suppose we declare the real numbers to be all strings that begin with the $+$ or $-$ sign and continue with a finite or infinite number of digits. Each string has an embedded decimal point. Calling that construction of the real numbers a discovery would give undeserved recognition to a trivial idea.

We could use the strings in exactly the same way as we employ the real numbers defined by the Dedekind cut described in Chapter 2. Why didn't Dedekind offer that simple string definition?

The answer is that he wanted to show that the well-understood rational numbers—which are ratios of integers—could be used to construct all irrational numbers. In his words,

"Just as the negative integers and fractions of integers are produced by free creation, and just as the laws of computations must and can be based on the computations for natural numbers, in like manner one has to strive that the irrational numbers as well are defined entirely via the rational numbers. Only the How? remains the question."[518]

The Dedekind cut supplies the How.

Similar stories can be told about the complex, algebraic, and transcendental numbers covered in Chapter 2.

Gauss likely had the history of all these numbers in mind when he wrote, "[N]umber is purely a product of our mind."[519]

Remember the extraordinary battle of Leibniz and Newton[520] about the question who had developed calculus first? We look once more at that conflict.

Calculus

Consider the following made-up and outlandish—or rather island-ish—story. Suppose Leibniz and Newton are explorers eager to discover new lands. They set out to sea, and after some weeks of travel across the Atlantic Ocean, Newton discovers an island.

A bit later, Leibniz also chances upon an island. Excited, both sail home and tell about the discovery. Now neither of them gets an award or monetary payment for the discovery.

At first, they mail each other information about their finds. The descriptions of the two islands differ substantially, so for a while they think that they have discovered different islands. But then going into details, they realize that they had found the same island.

How would they react? As life-long enemies, each of them claiming that the other one was cheating? This seems unlikely. After all, the debate would be who had *seen* the island first.

So they probably would say, "It's amazing; we sailed across large unexplored parts of the Atlantic Ocean quite unawares of each other and discovered the same island."

Compare this with the actual events. Newton and Leibniz worked very hard for several years to develop and refine the theory of calculus.[521]

Both succeeded where earlier mathematicians had only produced incomplete results. There was no financial reward whatever, only recognition that this difficult problem had finally been solved. That acknowledgement alone motivated them to defend their claims with relentless effort.

Given this scenario, what is more reasonable to assume: That they valued their work as the creation of a new method; or that they viewed it like the discovery of an island?

It is generally agreed that Bürgi and Napier independently came up with the idea of logarithm.[522] We examine that development from Bürgi's viewpoint.

Computing with Logarithms

In mid-16th century, Stifel included the following two rows in a treatise on arithmetic:[523]

-3	-2	-1	0	1	2	3
$\frac{1}{8}$	$\frac{1}{4}$	$\frac{1}{2}$	1	2	4	8

He called the numbers in the top row *exponents*. When 2 is raised to the power of one such number, then the entry right below displays the result.

Stifel made the following crucial observation: When two numbers in the bottom row are multiplied, then the exponent of the resulting number is the sum of the exponents of the two numbers. In modern notation, for any m and n, $2^m \cdot 2^n = 2^{m+n}$.

Let's imagine the moment where Bürgi sees Stifel's two rows. He thinks, "The number 2 induces this behavior. But wait a moment. This may have nothing to do with the number 2. This may be true for *any* number a. Is then—in modern notation—$a^m \cdot a^n = a^{m+n}$?"

After a moment's reflection, he realizes, "Of course, this is implied by the rules of multiplication." Is conjecturing that generalization and then proving it a discovery or a new idea?

Let's consider another aspect. When Bürgi published his table of logarithms in 1620, he wanted to display on the title page the gist of the idea. He came up with a circular pattern that lists every 500th entry of the table. He must have been pleased that he found this compact and visually impressive display.

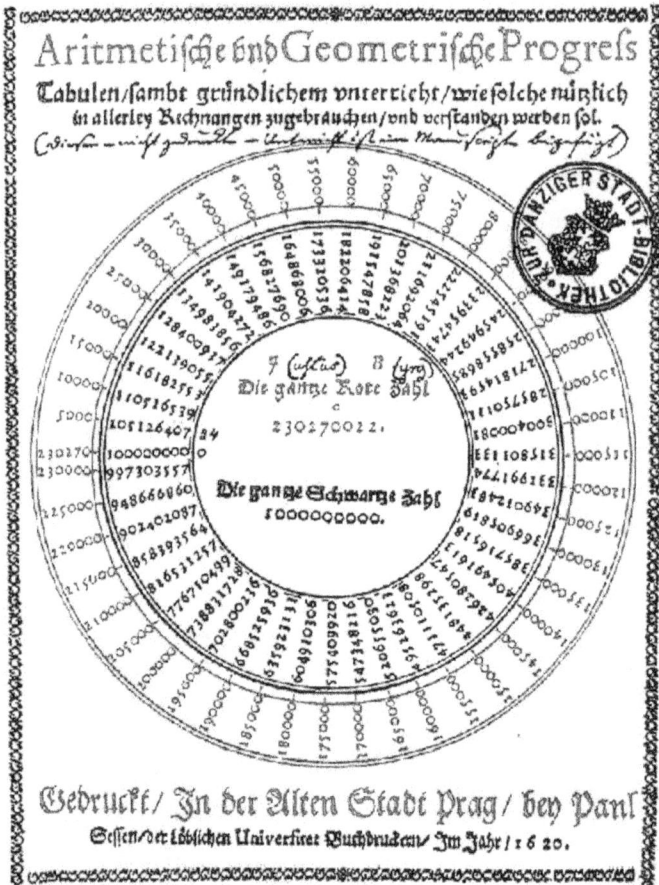

Title page of Jost Bürgi's Table of Logarithms, 1620. The page lists every 500th entry of the table. The numbers are in the inner ring, and the logarithms in the outer ring.[524]

Suppose Bürgi had combined the inner ring of numbers of the title page with a larger copy as shown on the next page. Effectively, he would have created the circular slide rule.[525]

Instead, William Oughtred (1574–1660), an Anglican minister and mathematician, invented the slide rule two years later, in 1622, and the circular slide rule after another ten years, in 1632.[526]

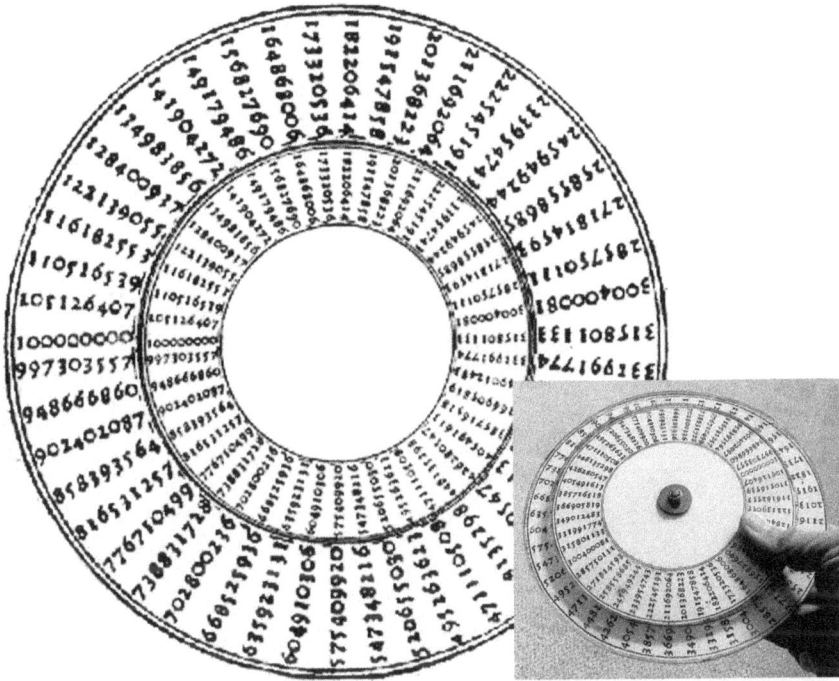

Circular slide rule assembled from two copies of the inner ring of numbers of Jost Bürgi's Table of Logarithms. Inset: implementation using plastic disks.[527]

Let's imagine the following scenario, for which we have no proof that it ever occurred. Oughtred sees Bürgi's title page.

Immediately he thinks, "Now I know how to construct a terrific device for computation. Let's call it the circular slide rule."

Would Oughtred thus invent the circular slide rule, or would it be a discovery?

As described in Chapter 3, Euler's idea of function opened a flood of developments where many types of functions were defined and investigated.

William Oughtred, by Wenceslaus Hollar.[528]

For example, Turing defined computability of functions and showed that some functions, in particular, the halting function, were not computable; see Chapter 3.

We look at a related aspect.

Knowledge of Function Values

Let's expand the concept of computability to the question whether one can ever *know* all values of a function, in the sense that we know that humans generally have a head, two arms, and two legs.

Of course, the values of a function can only be known if they exist. Is such existence guaranteed for any well-defined function when discovery of mathematics is assumed?

Let's look at an example. Consider a function F where the input is any set S. The output is 1 if the set S can be rearranged in some order so that every nonempty subset of S has a smallest element. The output is 0 if this cannot be done.[529]

We can think of a number of questions about the values of F. For example, are there input sets S for the function F so that the output value is 0? Or, more basically, what is the value of F when the input is the set of real numbers?

The fact is that these questions cannot be answered unless sophisticated machinery such as the well-ordering theorem or the equivalent axiom of choice[530] is applied.[531]

Indeed, there is no a priori information about all output values of the function F, even though the definition of the function is clear and unambiguous.

Instead, we create an answer once we adopt or reject certain axioms. Can this be reconciled with the concept of discovery of existing things or facts?

The next section looks at two key advances for the integration of functions.

Integration

In the 19th and beginning 20th century, Riemann and Lebesgue describe two different methods for computing the area under a function.[532]

Let's move both events to an earlier time; say 1750. Riemann is happy since he finally has managed to place the computation of area under a function on a solid foundation.

His method slices the area into vertical strips, computes the area of each strip, then adds up these values to the total area. The arguments are clear and complete.

Suppose he is visited by Lebesgue, who excitedly tells him of a new way to compute the area under a function.

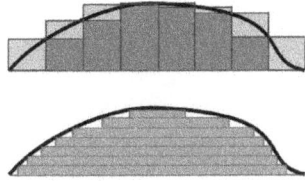

Top: Riemann integration.
Bottom: Lebesgue integration.[533]

The new method slices the area into horizontal strips, then computes the area of each strip by examining how their length is implied by the function shape. He adds up these areas to obtain the total area.

Let's listen to a fictitious discussion.

Riemann: "Your new method is interesting, but complicated. It may be quite difficult to compute the length of each horizontal strip. In contrast, my method only needs one function value for each vertical strip."

Lebesgue: "Yes, you are right. But my method readily handles functions that contain jumps without additional difficulty."

R.: "But the functions of interest to us may jump only a few times, comparatively speaking, and that is no problem for my method."

L.: (admitting defeat): "Well, I guess you are right. I just thought it was an interesting idea."

R.: "Yes, but it seems that it makes integration more complicated than it needs to be."

And so Lebesgue's idea would have ended up in the dustbin of discarded mathematics. Why then did Lebesgue's method—created around 1900 and not in 1750 as assumed above—become famous?

The above assertion by Riemann about jumps of functions correctly reflects the general thinking of mathematicians around 1750. That notion was destroyed in the 19th century, for example, by the function $f(x)$ defined on the interval from 0 to 1 that is equal to 1 if x is rational and equal to 0 if x is irrational. The area under that function cannot be computed with Riemann's method, but is successfully determined by Lebesgue's scheme.[534]

The point of this fictitious conversation is that a brilliant mathematical idea may be rejected if it occurs at the wrong time. Put differently, when Lebesgue came up with his method of integration around 1900, he considered how it fit into the then current understanding of functions. In fact, most likely his research was motivated by the inability of Riemann's method to handle recently constructed functions.

Can an understanding of time-dependent mathematical importance be discovered in an eternal repository?

The concepts of infinity and infinitesimals occupied mathematicians for centuries. Cantor was the first mathematician to clarify them.[535] We examine a fictitious exchange between him and a novice in mathematics.

A Claim of Infinities

Cantor is taking a walk in the park. A novice introduces himself and explains his work.

Novice: "I have a terrific insight into the infinities you have created. You say that the set of integers is a smallest infinite set, and that the set of real numbers is a next larger set."

Cantor: "That is my conjecture, and I firmly believe it to be true."[536]

N.: "Well, I have an alternate conjecture: There are an infinite number of sets whose sizes are all different and lie strictly between the size of the set of integers and the size of the set of reals."

C.: "Really? How can you possibly claim this?"

N.: "Well, I do. It makes sense, in the same way that there are an infinite number of rational numbers that lie between 0 and 1."

C.: "Do you have any evidence for this claim?"

N.: "Not yet."

At which point Cantor says, "Good-bye," and returns to his office.

Doing mathematics requires a large degree of discipline: Absent that, the effort is dissipated in wild conjectures and never leads to anything. Thus, a focus on hard problems must be tempered by an intuitive, personal judgment that lies outside the domain of techniques and proofs. Can such judgment be discovered?

The next three sections investigate scenarios of conjectures, proofs, and theorems.

Conjectures

Fermat conjectured[537] that, for $n = 1, 2, 3, \ldots$, the natural numbers $F_n = 2^{(2^n)} + 1$ are prime numbers. That is, each of them cannot be produced by multiplication of some natural numbers except for the trivial multiplication involving the factor 1.

The conjecture seemed well justified: The numbers F^n up to $n = 4$ are 3, 5, 17, 257, and 65537, all of which are prime.

But Euler showed that the next number, F_5, satisfies

$$F_5 = 2^{(2^5)} + 1 = 2^{32} + 1 = 4294967297 = 641 \cdot 6700417$$

which proves it to be not a prime.[538] Thus, Fermat's conjecture was incorrect. The F_n which indeed are prime are now called *Fermat primes*. It is not known whether there are any additional Fermat primes beyond $F_0, F_1, \ldots, F_4$.

Did Fermat discover the incorrect conjecture or did he create it? What would the answer be if the conjecture had been correct? Is the currently undecided conjecture "$F_0, F_1, \ldots, F_4$ are the only Fermat primes" created or discovered?

Proponents of mathematical discovery may be tempted to claim that the postulated realm of mathematical results does not contain erroneous conjectures. Assume this is true. Then the statement "It is conjectured that $1 + 2 = 5$" cannot be discovered. But the correct mathematical statement "The following conjecture is true or false: $1 + 2 = 5$" would be in the realm of mathematical results.

The latter construction can be applied to any false conjecture. Hence, effectively all false mathematical statements, preceded by "The following conjecture is true or false," would be in the realm of mathematical results.

Proofs

Consider the following claim by Cantor:[539] The set of natural numbers is an infinite set of smallest size. An equivalent claim is: Every infinite set contains, up to a relabeling of the element names, the set of natural numbers as a subset.

Cantor's proof may be rephrased as follows: Given an infinite set S, initialize N as the empty set. Remove one element from S, and place it as 1 into N. There still are an infinite number of elements left in S, so now remove another element from S, and add it as 2 to N. Continuing in this fashion, we eventually obtain $1, 2, 3, \ldots$ for N, which thus has become the set of natural numbers. Hence, the original set S essentially contains N and thus must be at least as large as N.

These arguments seemingly constitute a solid proof of the claim. But they do not. The defect of the proof can only be remedied by some additional assumption—for example, by assuming the axiom of countable choice.[540]

Did Cantor create or discover the flawed proof? A vote for discovery seems like a poor choice, since flawed statements then occur in the assumed realm of mathematical results—a strange notion for a place of perfection.

Consider the following situation. We have worked for years and have produced an ingenious proof for a very important theorem. We proudly show it to a friend. Upon closer examination, she finds a small error. Thus, the proof is wrong and cannot have been discovered. Indeed, we created it.

Looking closer, we quickly come up with a remedy. Was the corrected proof discovered? If so, we *created* the bulk of the proof, and *discovered* just a trivial modification.

Fermat's last theorem, claimed by Fermat in 1637 and finally proved by Wiles in 1995—see Chapter 6—says that the equation $x^n + y^n = z^n$ has no solution with integers x, y, and z if $n > 2$. Let's denote that famous theorem by *FLT*.

Prior to Wiles's proof, many flawed proofs of *FLT* had been published. Surely these failed proofs had not been discovered. Suppose that, taken together, they did provide Wiles considerable insight into the problem, in the sense that he could exclude broad areas of mathematics where a proof *couldn't* be found.

Was that insight discovered, even though it was gained from erroneous proofs that were not discovered?

Let's turn to theorems.

Theorems

We use *FLT* again. Let's reformulate it as declaration D:

"The following statement is true: The equation $x^n + y^n = z^n$ has no solution with integer x, y, and z if $n > 2$."

One of the arguments in favor of discovery of D is the following. Once we know that D has been shown to be valid by Wiles, we also know that it was valid at any earlier point in time. And thus—it is argued—statement D must reside in some realm outside the natural world.[541]

But is it a correct argument, or a nonsensical philosophical claim? On the surface, the reasoning seems irrefutable. But we know from Wittgenstein's image of the fly trapped in the fly bottle that seemingly irrefutable philosophical statements need not be correct.

So, following Wittgenstein's recipe for escaping the fly bottle, we look at the problem from different angles. Specifically, we view the assumed prior existence of statement D as some sort of time travel where D moves back in time from the point of Wiles's proof of validity. During that movement, D is always valid.

During that backward journey, we move past Fermat—the first human to write down the theorem—and stop momentarily at the time of Archimedes. We show him D.

Since Archimedes doesn't know the notation of exponents created by Descartes, he isn't able to interpret D. But we can rewrite the statement so that it doesn't involve any exponents and instead just mentions the natural numbers and their addition and multiplication. So Archimedes understands the modified statement. He might even try to prove it like so many other mathematicians.[542]

Let's go back in time much further and stop at a point tens of thousands of years ago. Piles of pebbles are used to account for quantities, but the natural numbers characterizing the sizes of piles haven't been created yet. Of course, the concept of addition and multiplication of natural numbers doesn't exist either. The situation is sketched at the beginning of this chapter.

So what does it mean when somebody declares that statement D is valid at that time? Surely nobody living at that time can understand it, and we cannot rewrite it to make it understandable since the natural numbers and their arithmetic operations don't exist yet!

The situation is even worse if the reader has accepted the earlier arguments showing that all numbers and their basic arithmetic operations are man-made. In that case, a mathematical statement, here D, is claimed to exist at a point in time where the constituents of the statement do not exist as yet. A very odd result indeed.

There is a way out of the latter predicament: First, one declares that concepts and axioms of mathematics are created.

Second, one claims that, as soon as such a creative step has been taken, all theorems that can possibly be derived from all now-available concepts or axioms, come immediately into existence. Subsequent derivation of any such theorem by humans is then always discovery and not creation.

So for the case at hand, statement D comes into existence when, thousands of years ago, natural numbers and their addition and multiplication are first defined.

For the moment, let's ignore this revised discovery claim; we return to it later in this chapter. Instead, we look at another scenario that also involves time travel of mathematical statements.

A strong argument in favor of mathematical discovery is the *unreasonable effectiveness* with which mathematics explains the natural world.[543] To be sure, there isn't uniform agreement on this conclusion, and in a play on words, *reasonable ineffectiveness* of mathematics has also been claimed.[544] We take up this issue in Chapter 11.

For the discussion to follow, we assume—motivated by the claim of unreasonable effectiveness—that mathematics is part of nature and thus is discovered just as rules involving natural phenomena are discovered.

Under that assumption, Einstein discovered the theory of relativity and expressed it in appropriate mathematical formulas. This discovery includes the following statement:[545]

"$E = m \cdot c^2$ correctly describes the relationship between energy E, mass m, and the speed of light c."

Michelangelo (1475–1564), the famous Renaissance sculptor, painter, and architect created the *David* statue in 1504. We place the statue into the 3-dimensional Cartesian coordinate system[546] and define a function $F(x, y, z)$ that has value 0 if the point (x, y, z) is outside the statue, and to have value 1 otherwise. This definition is not precise: As we go down in scale to the level of molecules, it is not clear whether a given point is outside the statue or not.

Michelangelo, by Daniele da Volterra, 1544, detail.[547]
Left: David, by Michelangelo, 1504.[548]

We remedy this defect by selecting a discrete grid of sufficiently small stepsize, say 0.00001 inches, and then define the function for those grid points. When there is doubt whether or not a grid point is outside the statue, we simply declare it to be outside.

The error introduced by this rule is far below the surface roughness still left after Michelangelo polished the statue. Thus, for all practical purposes, the function F correctly represents *David*.

When Michelangelo completed the statue in 1504, the function F was well defined. So, the following statement M was valid:

"The function F [explicitly defined with the values of all coordinate points] completely specifies Michelangelo's *David*."

This is a correct mathematical statement where a mathematical object, a function, specifies something about the world, just as Einstein's mathematical object, an equation, says something about the world.

Imagine time travel involving statement M, say back to 500 BCE. But now the statement takes on a strange meaning: Since the function F exists in 500 BCE, *David* is already defined in 500 BCE.

As a consequence, Michelangelo can only discover, and not create, *David* in 1504, 2,000 years later. Indeed, he must take great care that he achieves the shape of the sculpture implied by the function!

This argument is bizarre, to say the least. The strangeness stems from the time-travel aspect made possible by the assumed time-independent existence of mathematical formulas and statements.

This strangeness of time travel is well known in science fiction. Indeed, the odd logic of such travel pleasantly tickles our brain even though, or maybe because, we must suspend everyday thinking. But do we want to introduce such thinking into our concepts about mathematics and the world?

In a worldwide effort lasting decades, sculptures and architectural wonders such as cathedrals and temples have been digitized and thus effectively recorded as mathematical functions. According to the above arguments, each of these marvels of human ingenuity and skill was discovered and not created.

True but Unprovable Statements

There are mathematical statements that are true but cannot be proved within the framework in which the statements have been defined. Gödel assembled such true but unprovable statements for all formal mathematical systems that contain a certain amount of elementary arithmetic when he established the first incompleteness theorem.[549]

It is tempting to claim that such true and unprovable statements have an a priori existence independent of Gödel's mathematical investigation. Thus, these statements seemingly are discovered.

The key for resolving the claim lies in the interpretation of the word "true." In this case, it means that the statements can be proved to be valid using a framework that is different from the one used for the definition of the statements.

Given that insight, we see that the arguments for discovery involve the following, by now familiar, reasoning:

Consider any statement of the type, "Here is a claim that cannot be proved within its framework of definition, but can be proved outside that framework: [explicit listing of the claim]."

We prove the statement to be valid. We conclude that it was valid since time immemorial, and hence has existed in some realm. Based on that conclusion, we say that we discovered the statement.

Actually, it looks just like the case of theorems of the preceding section, doesn't it? In this case, we have a theorem claiming that a statement is not provable within its framework of definition, but is provable in another framework.

The discussion so far may give the impression that mathematical results are mostly produced by deduction from axioms. The outstanding philosopher Imre Lakatos (1922–1974) assembled brilliant arguments in *Proofs and Refutations*[551] that show this viewpoint to be quite wrong.

The next section summarizes the key ideas.

Imre Lakatos, ca. 1960.[550]

Cycles of Flawed Theorems and Proofs

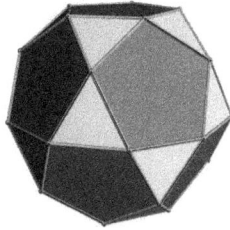

Icosidodecahedron.[552]

In 1758, Euler conjectured the following formula for *polyhedra*, which at the time were defined to be 3-dimensional bodies with flat boundary surfaces:[555]

Let V be the number of corner points of a polyhedron, F be the number of flat surfaces or facets, and E be the number of edges. Then Euler's formula is $V + F - E = 2$. He tested the formula for a number of polyhedra and proposed a proof that later was found to be defective. Indeed, the icosidodecahedron and stellated octahedron satisfy the formula, but the hexagonal torus, with $V + F - E = 24 + 24 - 48 = 0$, does not.

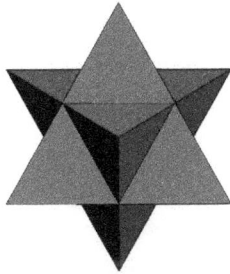

Stellated octahedron.[553]

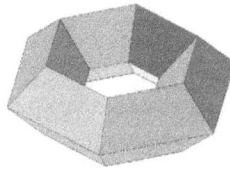

Hexagonal torus.[554]

Over the next 150 years, a number of mathematicians came up with proofs, counterexamples, further theorems and proofs, yet more counterexamples, and on and on.

Lakatos imagines a class of students exploring that history with the help of a teacher. In extensive discussions, they investigate Euler's formula and its various proofs, counterexamples, and modifications. The teacher cleverly guides the students so that they experience the events in historical order.

In Wittgenstein's terminology, Lakatos creates a language game about the history of Euler's formula. He operates that game to recreate that history in a classroom setting.

The discussions in the class—or, as Wittgenstein would view it, the operation of the language game—demonstrate that the traditional viewpoint of the mathematical process is wrong: It isn't just definition of theorems followed by verification via deductive proofs or annihilation via counterexamples.

In fact, the students of the class gradually recognize that counterexamples do not simply destroy theorems, but can lead to further areas of investigation, and that flawed proofs can be valuable steps toward deeper insight and new concepts.

Unfortunately, we cannot amplify the above summary with citations of Lakatos's book. The dialogue of students and teachers takes up 52 pages, including historical notes, and is simply too detailed and extensive to be quoted here. But we do encourage the reader to obtain the book and experience the brilliance of Lakatos's arguments.

Lakatos goes so far as to claim that the traditional representation of mathematical development isn't just wrong from a philosophical viewpoint, but actually inhibits creative activity by painting a wrong picture of mathematics in the minds of students.

It seems difficult to reconcile the insight produced by Lakatos's language game with the notion of discovery of already existing mathematical results.

After all, the history of the Euler formula is littered with incorrect claims and proofs. Surely those wrong results were not discovered. But, they provided the insight required for mathematicians to resolve the difficulties connected with that formula.[556]

In the next section, we compare developments of music with those of mathematics.

Music

In a long evolutionary and at times revolutionary process, human creativity produced various forms of art. The accomplishments include elaborate plays and novels, lifelike sculptures and complex paintings upon which we gaze with awe, and music compositions performed by large orchestras.

Earlier, we cited some of these accomplishments to get insight into the development of mathematics. We now turn to music for the same purpose.

The discussion simplifies the development of music over thousands of years to just a few events. We place these milestones of music side by side with corresponding mathematical accomplishments.

We assume that all mathematical results are discovered and not created. That classification also applies to early developments in music; for example, identifying a birdsong can be viewed as discovery. But beginning with the definition of notes, the developments in music are no longer discoveries, but involve creation.

One may debate which milestones of music constitute the exact transition from discovery to creation. Regardless of the choice, we can compare that transition point with the corresponding developmental step of mathematics and ask, "Why do we declare a transition from discovery to creation for music while we postulate discovery for the corresponding step in mathematics?"

In Wittgenstein's terminology,[557] the comparison process is a language game where we explore two meanings of the word "development": discovery and creation.

As we shall see, the language game brings a conflict to the surface that can only be resolved by either declaring all developments of music to be discovery, or by abandoning the assumption that all of mathematics has been discovered. Wouldn't the former conclusion be objectionable to almost everyone?

Below, the milestones are assembled in two columns, where the left-hand side contains the events for music and the right-hand side has the corresponding events for mathematics.

Development of Music . . .	*. . . and of Mathematics*
1. Various waves exist in nature (discovery).	1. The world contains many different things.
2. Human ears perceive and classify sound waves (discovery, partially creation).	2. Define classes, such as the class of apples.
3. Select particular sound waves. Define notes C, D, E, ..., C♭, C♯, ... (creation).	3. Start counting. Introduce natural numbers, integers, rational numbers, real numbers.
4. Organize notes for composition (creation).	4. Define basic arithmetic.
5. Design and build music instruments (creation).	5. Develop rules of logic and deduction.
6. Bach composes Brandenburg Concerto No. 3 (creation).	6. Bürgi and Napier develop logarithm and related tables.

The column for music contains for each step in parentheses a seemingly reasonable classification as discovery or creation. For mathematics, we omit such explicit classification since discovery is assumed throughout.

If the reader disagrees with the above discovery/creation classification for music, it can be changed as may seem appropriate. But we anticipate that there will still be agreement that waves in nature were discovered, and that Bach's concerto was created. Thus, there will always be a transition from discovery to creation for music, while mathematics by assumption always involves discovery.

Look at any step with a classification conflict between music and mathematics. The conflict occurs above in Step 3, where humans define notes in music as well as numbers in mathematics. Why would the case of notes be creation and that of numbers discovery?

A proponent of discovery of mathematics may claim that the above discussion is irrelevant: Music and mathematics simply are so different that the above comparison is ill-advised.

In particular, mathematics operates under the concept of eternal correctness, while music allows virtually any sequence of notes.

But how do the axiom of choice and the continuum hypothesis fit into the claimed eternal correctness? We may assume them or reject them, and in each case get quite different conclusions.[560]

And how does that argument apply to the composition *A Musical Joke* (Ein musikalischer Spaß)[561] by Wolfgang Amadeus Mozart (1756–1791)?

It purposely contains composition errors and typical performance gaffes of dilettante musicians. Listening to a performance, we recognize that the composition is boring and repetitive.

Johann Sebastian Bach, by Elias Gottlob Haussmann, 1746.[558]

Wolfgang Amadeus Mozart, posthumous painting by Barbara Krafft, 1819.[559]

We also recognize the instances where notes are improperly played, in particular by the horns. That is to say, there are complex rules for composition that must be observed if we are to find a piece interesting, and we notice when a performer hits wrong notes specified by Mozart.

Here is another objection to the comparison of music with mathematics: There are many more choices open to the composer than to the mathematician, since mathematical arguments are rigidly constrained by rules of logic. But those rigid rules admit huge collections of theorems, indeed gazillions of them, that we discard and declare to be trivial or uninteresting, or that we replace by a few axioms.

For example, consider the theorem that, for any two integers a and b, there is an integer c so that $a + b = c$. In *Principia Mathematica*, it takes Whitehead and Russell 379 pages to prove a logic statement[562] equivalent to the arithmetic result $1 + 1 = 2$.

Similarly, we could spend our lifetime proving additional theorems for the integers: $1 + 2 = 3$, $1 + 3 = 4$, and so on. We pass over that infinite collection of theorems as uninteresting and instead invent, say, the axioms for addition of the integers. Then in one fell swoop, all those boring theorems disappear, or rather, no longer need to be proved. We do this lots of times, replacing a huge collection of theorems with a few simple axioms.

So there are many, many theorems in mathematics, but we ignore virtually all of them and focus on the comparatively few we feel are interesting, just as a composer considers only note sequences that appear interesting.

Finally, a proponent of mathematical discovery may argue that the composition rules of music have changed over time. But so have the axioms acceptable in mathematics; see Chapters 3, 4, and 6.

Created Concepts, Discovered Consequences

Earlier in this chapter, the following claim came up but was not addressed: Mathematical concepts are created, but all consequences provable from these concepts are discovered. We now look into that claim.

In some sense, the creation of a new concept *causes* all possible future conclusions derivable from the totality of now-available concepts to come immediately into existence. Subsequent human derivation of such conclusions is then discovery and not creation.

The assumed process of creation of concepts and discovery of consequences raises a number of questions. Let's look at some examples.

Does the postulated process take place when the concept is flawed and the derivation of results is not correct according to present-day standards—as happened in the case of the concept of infinitesimal quantities and their use by Wallis, Leibniz, and Newton?[563]

What occurs when a concept is introduced and later abandoned—such as the introduction of intuitionism[564] in the 1910s by Brouwer and its abandonment by the mathematical community in the 1930s?

How do we interpret the situation when a result has been proved, the proof is found to be defective and cannot be repaired within the existing framework, but is corrected *afterward* by introduction of a new concept? An example is Cantor's erroneous proof of the theorem that the set of natural numbers is a smallest infinite set, and the subsequent remedy by creation of the axiom of countable choice.[565]

Which results come into existence when a concept is vaguely or incompletely specified, or when various opinions exist about it—as is the case of the concept of finitary proof, introduced by Hilbert and as of 2016 still not completely specified?[566]

The search for answers to such questions seemingly leads to a thicket of philosophical arguments, so we shall not attempt such investigation.

Instead, we explore implications of the claim of creation of concepts and discovery of conclusions using the earlier specified language game involving music. The early steps of music were discovery and the latter ones creation, while all steps of mathematics were assumed to be discovery.

Let us change this so that the assumption of this section is satisfied. That is, the early steps of mathematics forming concepts are now creation, while the subsequent steps deriving conclusions are discovery.

The reader likely will agree that Step 5 of mathematics, "Develop rules of logic and deduction," produces consequences of fundamental concepts of logic and thus by assumption is discovery. In contrast, the related Step 5 for music, "Design and build music instruments," is declared to be creation as before. Why would we have such clashing classifications of similar endeavors?

William Shakespeare, by Martin Droeshout, 1623.[567]

Let's try an alternate approach. That is, we translate the claim of creation of concepts and discovery of results in mathematics directly to the world of music. We then get the following postulate: After mankind created music notes, all compositions—including Bach's Brandenburg Concerto No. 3—were discovered.

Similarly, we arrive at the following postulates: After the sculptors of ancient Greek had created the methods for working with marble around 500–400 BCE,[569] all marble sculptures produced by these methods over the ensuing centuries—including Michelangelo's David statue—were discovered.

L'ultimo bacio dato a Giulietta da Romeo (Last kiss given to Juliet by Romeo), by Francesco Hayez, 1823, detail.[568]

And finally, after mankind had created various forms of writing, all written documents—including Shakespeare's famous play "Romeo and Juliet"—were discovered.

Of course, we soundly reject these postulates. So why would we accept the corresponding claim for mathematics?

Summary

We have approached the question of creation versus discovery of mathematics from various angles, as suggested by Wittgenstein for the resolution of philosophical questions. Each time, the assumption of discovery resulted in odd if not outright bizarre consequences. This insight lends strong support to the claim that mathematics is created.

The next two chapters provide additional arguments buttressing the conclusion of a created mathematics. The chapters examine the following two claims. First, that mathematics is unusually effective in representing models of the world, and second, that mankind cannot possibly live without mathematics. These assertions have been viewed as partial proof of discovery of mathematics. However, careful examination reveals that they are unjustified.

11

Effectiveness of Mathematics

Today, mathematical methods and approaches assist with almost every human activity. Evidently, mathematics gives us deeper insight, guides us toward good decisions, results in accurate predictions about the future, and generally helps us understand the world.

How is it possible that mathematics can be used in such a profound and universal way? Put differently, why is mathematics so effective?

For any person believing that mathematics is part of the world and thus discovered, it isn't surprising that mathematical concepts such as formulas and algorithms can be used to understand the world. In fact, such a person is tempted to claim the following: The incredible effectiveness of mathematics for understanding the world is *proof* that mathematics is part of the world and thus is discovered and not created.

This argument seems utterly convincing and thus places the hypothesis of a created mathematics in serious doubt. But is the argument correct? This chapter investigates that question.

For a start, let's consider the following question: How does nature work? The answer for that question shifted considerably over thousands of years, moving from simplistic a priori claims to carefully worked-out postulates based on detailed observations.[570]

These claims and postulates could be understood by almost everybody right up to the beginning of the 19th century. But then the statements about the world became more and more complex and by now are so abstract that only the specialist fully understands the claimed relationships. Examples are the general theory of relativity and quantum physics.[571]

There are admirable attempts to clarify matters for the layperson.[572] Interestingly enough, we not only get a better understanding of science from such enlightening material, but also are forced to draw the following conclusion: We no longer have direct insight into the world in the sense that we know intuitively what does or does not exist.

Instead, we have elaborate models about the world that have been confirmed by numerous experiments. We then act as if the models were the actual world. This viewpoint, called *model-dependent realism*,[573] seems to be the only reasonable basis for interacting with the modern world.

For example, the Global Positioning System (GPS) uses clocks in satellites to time signals and thus compute distances.

The speed of the clocks depends on the gravitational force exerted by the earth and the velocity of the satellites. Suitable adjustments of the clocks are computed using the gen-

GPS satellite in Earth orbit, artist's conception.[574]

eral theory of relativity.[575] GPS *works* in the sense that we can compute positions on earth with astonishing accuracy. But we have no clue whether we are using the right model for the computations.

In fact, we shouldn't use the term "right model," since we don't know what the world is actually like. Even the term "is actually like" is misleading. It presumes that there is a well-structured world that we will eventually fully understand with sufficiently precise instruments and measurements. We have no evidence whatsoever that this will happen. Instead, we define models that work for us with sufficient accuracy.

If a model doesn't live up to that standard, we improve it or replace it with a different model. Note that "sufficient accuracy" is a relative term: It measures model performance using the results of yet another man-made model!

Thus, the question "How does nature work?" can only be answered with, "Fundamentally, we don't have a clue. We just build models that represent features of the world in a sufficiently precise manner. Then we act according to those models."

Things can go wrong with this approach. There are simple failures due to such models, such as drifting seasons caused by a poorly constructed calendar. But there are also catastrophic failures, such as the religion-based slaughter of millions of people because their model of a deity did not agree with some other model.

Given the concept of model-dependent realism, it is not difficult to see why mathematics appears to be so effective: Since the same people who developed concepts about the world also constructed the mathematics to represent those concepts, it's no wonder there is such agreement of mathematical models with concepts!

To back up this argument, we later look at an example where the concept of the universe changed over time, and where applicable mathematical models were modified accordingly.

Remember the earlier statement that effectiveness of mathematics may well establish mathematics to be part of nature and thus prove that mathematics is discovered and not created?

The fact that mankind developed both the models about the world and the mathematics for understanding them, constitutes a simpler explanation why mathematics is effective. Accordingly, effectiveness should not be used to argue for discovery of mathematics.

We look at a more fundamental problem faced by the claim of effectiveness.

A Disagreement over Mathematical Effectiveness

So far we have assumed that the following statement, made in the opening paragraph of this chapter, is correct:

"Today, mathematical methods and approaches assist with almost every human activity. Evidently, mathematics gives us deeper insight, guides us toward good decisions, results in accurate predictions about the future, and generally helps us understand the world."

But is mathematics actually that effective? Three papers have examined this question in detail.

The first paper, titled *The Unreasonable Effectiveness of Mathematics in the Natural Sciences*,[576] declares that mathematics is invented and in an almost miraculous way explains many results of the natural sciences.

The second paper reuses the first part of the above title, that is, *The Unreasonable Effectiveness of Mathematics*.[577] It looks at the effectiveness claim from various angles and offers several counterarguments. But in the end, it concludes that mathematics is indeed unreasonably effective.[578]

The title *The Reasonable Ineffectiveness of Mathematics*[579] of the third paper already indicates that it is opposed to the earlier conclusions. Indeed, it discusses a large number of examples to expose flaws in the earlier arguments.

Written from the viewpoint of an engineer, the third paper shows that many problems of modern science cannot be solved with existing mathematics. In particular, relationships we discern now by experiments are often so complex that we revert to simulation to understand processes or make decisions. Examples are complex engineering design problems and predictions made for weather or economic performance.[580]

How can one explain the existence of such diametrically opposed opinions? A reasonable answer is: The invention of the computer in the mid 20th century made massive calculations possible within decades. The first two papers, written in 1960 and 1980, did not anticipate that development.

In contrast, the third one, written in 2013, considered the tsunami of computational results that has changed mankind's view of the world.

At the same time, the demand for model precision has increased to levels never envisioned before. Many analytical mathematical models are no longer precise enough, and complex situations typically are evaluated now by computer simulation and other numerical processes.

We examine one situation in detail where over centuries the view of the world repeatedly changed, and where mathematical models were modified in tandem. But eventually computer simulation was, and still is, employed for precise results. The problem is the prediction of planetary movements in our solar system.

Antikythera Mechanism

Chapter 7 includes an early model of the movement of the planets: the Antikythera mechanism of the second century BCE, which implements a *geocentric* model, where the earth is placed at the center of the universe.

Analysis of the computing device has established that it was not very accurate. For example, the Mars pointer can be off by as much as 38 degrees. This is not due to inaccuracies of the gear ratios in the Antikythera mechanism, but reflects deficiencies of the underlying mathematical theory.[581]

Ptolemy: Geocentric Model

Claudius Ptolemy (100(?)–170(?)) set out to build a complete geocentric model.[582] For the sun and stars, this was not so difficult.

But the planets at times reverse course in *retrograde* motions[583] and thus seem to wander around, as is reflected in the word "planet," which in Greek means "wanderer."

Ptolemy represented the complex movements of each planet by a primary eccentric circular motion that was augmented by a smaller circular motion called an epicycle. But even such advanced modeling could not explain the actual behavior of the planets without significant errors.

Claudius Ptolemy.[584]

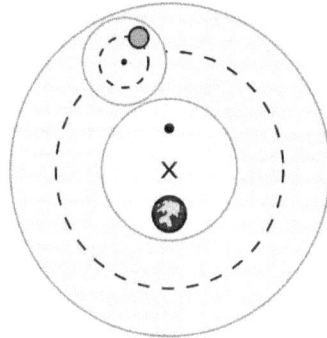

Copernicus: Heliocentric Model

In the first part of the 16th century, Copernicus moved away from the geocentric viewpoint and postulated that the sun was at the center of the universe and was circled by the earth and the planets.[586] See Chapter 4 for Galileo's role in this development.

Ptolemy's eccentric cycle centered at X plus epicycle centered on the eccentric cycle produce the orbit of the planet indicated by the small shaded circle. The earth is shown below the X with the continents.[585]

This *heliocentric* model was more precise, yet much simpler. But it did not fully explain the trajectories of the planets. Copernicus defined small epicycles for the planets to compensate for that shortcoming.

Kepler: Elliptic Orbits

Next, Johannes Kepler (1571–1630) postulated a heliocentric universe where the orbits of the planets are defined via nested *platonic solids*.

Each platonic solid looks like a cut diamond where all facets have the identical shape.

Due to this severe condition, there are just five platonic solids: tetrahedron, cube, octahedron, dodecahedron, and icosahedron.[588]

But even when eccentric circles and epicycles were used, the model never explained the movements of the planets with sufficient precision.[589]

Relying on data of the retrograde motion of Mars, Kepler used extensive calculations to eventually verify that the orbit of each planet around the sun is very well approximated by an *ellipse*, a curve known since antiquity.

An ellipse is defined by two *focal* points and the condition that the sum of the distances from any point on the curve to the two focal points is constant.[592]

In Kepler's model, the sun resides at one focal point of each planet's ellipse.

Newton: Gravity is Reason for Orbits

While Kepler's model explained *how* the planets are orbiting the sun, Newton determined *why*, using the concept of gravity.

Johannes Kepler.[587]

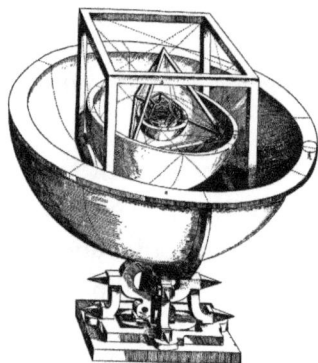

Kepler's model of the universe using nested platonic solids.[590]

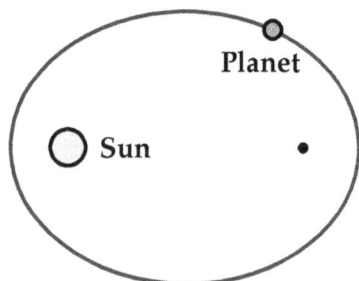

Kepler's elliptic orbit of a planet around the sun, which resides at one of the two focal points. The dot marks the second focal point.[591]

Newton's model is very general: It explains the speed with which an apple drops from a tree just as it describes how each planet orbits the sun.[593]

His formulas describe the interaction of one planet with the sun, when he should have considered the simultaneous interaction of all planets with the sun and each other. The *n-body problem*[594] demands mathematical description of the movement of *n* such bodies.

Newton was aware of this shortcoming and investigated the three-body problem—in particular the interaction of earth, moon, and sun. But he could not develop formulas describing the three-body case, let alone the general case of *n* bodies.

No wonder: In the 19th century, it was shown that the motion of three bodies in general is non-repeating, and that a compact mathematical description using algebraic expressions and integrals is not possible.[595] It was the first demonstration that the movement of the planets as postulated by the gravity model could not be represented by compact mathematical formulas.

Einstein: Theory of Relativity

Newton's model assumes that space extends in all directions and time flows uniformly. This postulate was reasonable given the accuracy of measurements of his period. In the 20th century, Einstein showed that this assumption about space and time is flawed. He then replaced Newton's model with the general theory of relativity, where gravity is represented by curvature of space.[596]

Today's Representation of Orbits

So how do we determine the movement of the planets today?

There are no compact formulas. Instead, the movements are determined by numerical methods.[597] Effectively, we simulate the movements on a computer, using mathematical postulates for the inter-action of the sun and the planets.

This approach works well for a limited time horizon, but supports only approximate description of the long-term evolution of the so-lar system.[598]

Let's summarize how effective mathematics has been in represent-ing mankind's theories on planetary movement.

Clearly, the Antikythera mechanism, Ptolemy's geocentric model using cycles and epicycles, Copernicus's much simpler heliocentric model, and Kepler's ellipse model demonstrate that over centuries man-made concepts could be encoded in analytical mathematical formulas.

Then Newton ran into the problem where he had an elaborate framework for planetary movements, but could not represent his ideas in compact formulas.

The goal of compact formulas became even more elusive with Ein-stein's general relativity model, and today we must be content with the fact that computer simulations determine planetary move-ments over a limited time horizon with high precision.

There are examples where analytical mathematical models have *always* failed. We will take a look at two of them: the predictions of weather and of economic performance.

Weather Prediction

Early attempts of weather prediction via analytical mathematical models failed badly. Eventually, the point was reached where weath-er was declared to be a *chaotic process* where small changes can trigger unexpectedly large consequences.[599]

Can you see what happened here? Mathematics was not able to predict weather. Declaring the weather to be chaotic made impre-cise predictions no longer the fault of mathematics.

There is an alternate explanation: Maybe analytical mathematical models simply cannot handle the complexity of weather phenomena.[600]

Indeed, weather forecasting now relies on elaborate computer-based models[601] that produce reasonably reliable short-term predictions consisting of a range of outcomes.

For example, the range of a hurricane's path can be predicted for a few days. When the forecast horizon is just a few hours, the process is called *nowcasting*.[602]

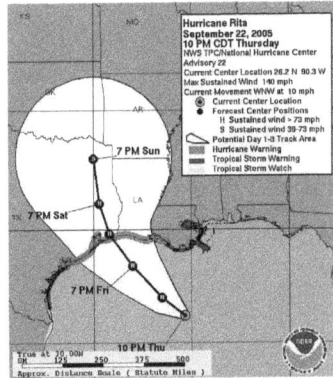

Three-day prediction of path for Hurricane Rita, 2005, by NOAA. The white area represents the predicted range of the path.[603]

Forecasting Economic Performance

Economic forecasts are another example where mathematics has failed to produce reliable results.[604]

For example, the world-wide recession that started in 2008—the worst since the Great Depression of the 1930s—was not predicted by most forecasters.[605]

Similarly to the weather case, the numerous failures of mathematical models have been explained by declaring economic processes to be chaotic.[606] Not everyone agrees with this assessment.[607]

Computer simulation and numerical methods have become key tools.[608] The success of nowcasting for weather prediction over a short horizon has spawned an analogous development in economics.

Here, nowcasting predictions are made not only for the short-term future, but also for the recent past where reliable data aren't yet available.[609]

21st Century: Dawn of Universal Computation

The move away from complete analytical models toward computational models has taken place in all sciences.

Indeed, if we were forced to characterize the 21st century using just one term, we would declare it to be the *dawn of universal computation*: No matter where we look, computers are carrying out ever more difficult tasks of evaluation, prediction, and decision making that cannot be handled directly with analytical mathematical models. Yes, some analytical models are used, but they mostly represent components in an overall computational approach.

An ever expanding use of increasingly powerful computers has also taken place *within* mathematics, indeed has led to a new research area called *experimental mathematics*. A summary of the goals and programs of this area is included in Chapter 13.

Summary

Not so long ago, it seemed that mathematical models represented the world with astonishing accuracy. One explanation was that mathematics is part of nature, leading to the conjecture that mathematics is discovered.

We have seen two arguments that mathematics does not provide the level of effectiveness once thought:

First, we have defined mathematics so that we can use it to model processes of nature defined by us. No wonder *our* mathematics can handle the analysis of *our* models.

Second, the accuracy of mathematical models has limits that we now overcome using unprecedented amounts of computation.

Thus, the effectiveness of mathematics cannot be used as an argument that mathematics is part of nature and can only be discovered.

It sometimes is argued that life is unthinkable without mathematics. For example, no matter where we look, there seem to be numbers: On a table, we see five apples, and in front of a house, two trees.

Indeed, numbers seem so important for human existence that we cannot possibly live without them.

If that argument is correct, the conclusion seems inescapable that at least some parts of mathematics are discovered. We examine this argument in the next chapter.

12

Life Without Mathematics

Mathematics is used everywhere. Statements such as "Why don't you come 15 minutes earlier," or "The store is open 24/7," or "Does this include the 8% sales tax?" show how tightly mathematics is woven into the daily life of modern mankind.

Indeed, the dependence on mathematics for even trivial exchanges, such as the request "Give me five minutes" and the accommodating response "You have ten," is so pervasive that human life without mathematics seems impossible.

Due to this extraordinary dependence, we seemingly are forced to conclude that mathematics is part of nature, and hence is discovered by the human mind and not created.

These arguments, convincing as they may seem, are wrong: Human life without *any* mathematics is possible. Proof is supplied by the Pirahã, a tribe of hunter-gatherers living in the jungle of the Amazon basin in Brazil.

The world of the Pirahã is dangerous: They are threatened on land and rivers by powerful and dangerous animals such as jaguars and snakes.[610]

In response, they have developed an extraordinary culture and language that allows them to not just cope with that world, but to thrive.

Overview of Brazil. The inset shows the Amazon region where the Pirahã people live.[611]

Let's first look at the Pirahã language.[612]

An Extraordinary Language

Pirahã is a special language that has just eight consonants and three vowels. Communication is possible in five ways.

1. *Regular speech*: uses consonants and vowels. Like a few other Amazonian tongues, there are male and female versions of spoken language: The women use one fewer consonant than the men do.

2. *Hum speech*: for privacy, like whispering in English. Can disguise one's identity. Mothers use it with children, or when mouth is full.

3. *Yell speech*: used during loud rain and thunder. Vowel "a" is used almost exclusively, and one or two consonants. Used to communicate over long distances, for example, across a wide river.

4. *Musical speech*: uses no consonants or vowels. Amplifies changes of pitch and rhythm of words and phrases.

5. *Whistle speech*: only used by males; for example, while hunting so that animals are not alarmed.

The Pirahã's communication via musical speech and whistling almost forces on us the conjecture that the songs of whales as well as the clicks and whistles of dolphins[613] may represent far more complex and sophisticated communication than assumed so far.

The Pirahã consider all forms of human discourse other than their own to be laughably inferior, and they are unique among Amazonian peoples in remaining monolingual.

They call their language "straight head" and label any other language with the derogatory expression "crooked head." Given the sophistication of their communication methods, the differentiation seems to be well justified.

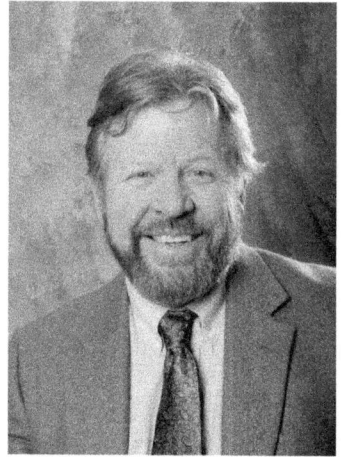

Daniel L. Everett.[614]

The language confounded early visitors. The first outsider to analyze and completely master the language was Daniel L. Everett (1951–).

Let's look at the language of the Pirahã in the context of their unique culture, which focuses on the present. Using currently favored terminology of the Western world, they "live in the moment."

Living Here and Now

The Pirahã have no past tense, no deep memory, no fixed color terms, no tradition of art or drawing, and no words for "all," "each," "every," "most," or "few"—terms of quantification believed by some linguists to be among the common building blocks of human cognition.

Part of the Pirahã's living-here-and-now is the following. If some-body makes the claim that something exists, the Pirahã ask if the person has *seen* that thing.

If the answer is "no," they ask if the person has learned about the claim from somebody who *has seen* the thing.

If the second question is also answered with "no," they simply do not believe the claim.

The reliance on actual witnesses for existence claims has blocked all efforts to convert the Pirahã to Christianity: Since there are no living persons who have seen Jesus or who have met a person who has done so, any story about Jesus is not believed.

The Pirahã language generally consists of short statements, each of which asserts a simple fact.

Indeed, Everett argues that the Pirahã language does not use *recursion*, defined in linguistics to be repeated application of rules that grow simple sentences to more complicated ones.

Here is an example of recursion. We begin with the sentence "The man sat on the chair." Then we expand this by recursion to "The man, who wore a hat, sat on the chair." Then, "The man, who wore a hat, sat on the red chair." The process could go on further for any number of steps, each time increasing the content and complexity of the sentence.

Everett's assertion of absence of re-cursion in the Pirahã language clashes with the claim by the famous linguist, philosopher, cognitive scientist, and historian Noam Chomsky (1928–) that recursion is the cornerstone of all hu-man languages.

Indeed, any language not using re-cursion supposedly cannot produce complex utterances of infinitely var-ied meaning.[616]

Noam Chomsky.[615]

There have been heated exchanges about presence or absence of recursion between Chomsky and researchers supporting him on one side and Everett on the other. Everett argues against the need for recursion using numerous examples of the Pirahã language.[617]

Viewing the controversy from a different vantage point, there are indications that recursion is not mandatory for rich human communication. Let's look at some arguments supporting that thought.

Meaning and Recursion

For the first argument, suppose Wittgenstein had been asked whether recursion was the core mechanism for profound human communication. With considerable confidence, we may assert that his answer would have been "no" for two reasons.

First, the claim that recursion induces complex meaning is an aspect of the picture theory of his Tractatus that he repudiated in the 1920s.

Second, Wittgenstein in his later work on language games includes the following statement:[618]

"It is easy to imagine a language consisting only of orders and reports in battle.—Or a language consisting only of questions and expressions for answering yes and no. And innumerable others.— And to imagine a language means to imagine a form of life."

Everett takes Wittgenstein's link of language and form of life further, saying that culture has an important role in the shaping of language.[619]

The second argument is based on a famous anecdote where Ernest Hemingway bet he could write a story with six words. He supposedly constructed such a story and collected on the bet.

There is doubt about the veracity of the tale,[620] but that doesn't matter here. What counts, is the story consisting of six words:

"For sale: Baby shoes. Never worn."

It is an example where no recursion is invoked, yet complex meaning unfolds in the reader's mind.

The third argument involves the two shortest, but likely fictional, telegrams.[621] Oscar Wilde supposedly sent the following telegram from Paris to his publisher in Britain to inquire how his newest book was doing in the market: "?" The publisher responded with "!" since sales of the book were doing very well.

The fourth argument is as follows. Suppose that recursion is indeed the key ingredient for creating profound meaning of sentences. Then one would expect that the analysis of sentence structure along the lines of recursion would allow computation of the meaning of sentences. Indeed, such analysis was carried out: *Transformational grammar*[622] supposedly allowed conversion of a given sentence with so-called *surface structure* to a new sentence revealing key relationships and dependencies through so-called *deep structure.*

The latter sentence was then alleged to allow analysis of the meaning of the original sentence. As a consequence, reliable translation of one language into another would be possible.

This approach did not succeed. Indeed, until recently, attempts at machine translation of one language to another largely have failed. But by now, in 2016, translated texts obtained via sophisticated software are useful but still far from perfect.[623]

How is it possible that we deduce complex meaning from a story of three sentences with two words each, or that we can understand the exchange between Oscar Wilde and his publisher? Or that machine translation is so difficult?

In the next chapter, we see an interesting explanation produced by brain science. The results described there also imply that recursion is not necessary for complex meaning.

We leave the topic of recursion and its role in language and come to the key section of this chapter, which concerns the absence of mathematics in the life of the Pirahã.

Absence of Mathematics

The Pirahã do not count and do not understand the concept of counting.[624] In fact, they do not have the number 1. For quantities, they have a word for "small." But it does not refer to a single item.

Everett tried for eight months to teach the Pirahã rudimentary mathematics, using Portuguese words for the numbers. After that time, not one Pirahã had learned to count to 10. None learned to add $3 + 1$ or even $1 + 1$. Only occasionally would some get the right answer.

How is this possible? Everett explains that, in the culture of the Pirahã, there is no need for numbers, and hence the language does not have words for them.[625]

For our purposes, the reason for the absence of numbers is not so important. What matters is that, for the discussion of creation versus discovery of mathematics, the Pirahã do not have numbers and do not care to learn about numbers.

Thus, one cannot argue for discovery of mathematics by saying that *all* humans use mathematics to function in the world and that therefore mathematics must be part of nature, to be discovered as needed.

Summary

We have met the Pirahã, an unusual people who do not use numbers at all. At the same time, they function very well in a hostile world that would quickly devour an unwary visitor.

So not knowing numbers is not a defect; the Pirahã simply do not need them in their lives.

Accordingly, it cannot be argued that mathematics must be part of nature since all humans use it.

As we have seen in Chapter 8, eminent mathematicians have made various conflicting claims concerning the creation or discovery of mathematics.

How is it possible that the human mind can come up with such contradictory assertions?

The next chapter attempts an answer based on the results of modern brain science.

13

Brain Science

Chapter 8 includes a number of conflicting claims by eminent mathematicians about creation or discovery of mathematics. How is it possible that the human mind not only produces such contradictory results, but defends each of them with vigor?

We don't see such a situation in the sciences. Yes, conflicting theories sometimes coexist in a science, but eventually one of them wins out. Later, the winner may be replaced by yet another theory.

For example, the claim that the sun was at the center of the universe had to compete for a while with the postulate that the earth was at the center. But then the former viewpoint won, only to be replaced in the 20th century by the all-encompassing Big Bang theory.[626]

It is tempting to resolve the question about the variety of conflicting philosophical claims about mathematics cited in Chapter 8 with additional philosophical arguments.

Given Wittgenstein's insight into the pitfalls of philosophy,[627] we can expect that additional philosophical arguments would just increase the fog of confusion.

Let's try an approach using science. Wittgenstein proposed the image of a fly in a fly bottle to depict how the brain may be trapped in a cycle of erroneous philosophical claims.[628] Can such faulty reasoning be traced back to a *structural feature* of the brain? If so, has that feature caused the cited conflicting claims about the origin of mathematics?

Not long ago, most questions about the brain's structure could not be answered. But since the 1990s brain science has undergone an upheaval where many prior results about the human brain have been proved wrong.

At the same time, many new results have been obtained in *systems neuroscience*, which studies the organization and workings of the brain using novel investigative techniques. Furthermore, in *molecular biology* new insights have been gained into the biochemistry of nerve cells and other cells in the central nervous system.[629]

In this chapter, we will use some of these results to develop a conjecture about the feature of the brain that makes production of contradictory philosophical claims possible.[630]

Note that we say "develop a conjecture about" and not "establish." The cautious tone reflects the fact that current knowledge about the brain's activities is still quite incomplete.

Nevertheless, we dare to use our hypothesis to explain not only why contradictory philosophical conclusions occur, but also why the language games invented by Wittgenstein for elimination of such confusion are so effective.

We begin with an attractive, but ultimately incorrect, assumption about how meaning of language can be computed. The postulated process parses a given sentence, identifies the logic connecting the parts, looks up the meaning of individual terms, and finally uses this information to compute the meaning of the sentence.

Wittgenstein developed the most sophisticated version of that explanation in his Tractatus;[631] it is now called the *picture theory*.[632]

Part of the picture theory is the postulate of *logical atomism*,[633] in which the world is thought to consist of ultimate facts that cannot be further subdivided or broken down.

These *atomic* facts are combined using mathematical logic to produce complex concepts.

Wittgenstein recognized in the second half of the 1920s that the picture theory was quite wrong. He then set out to develop methods that provide insight into the workings of language. His language games are a key tool for achieving that goal.[634]

So how do humans understand language, or more broadly, understand pictures, recognize emotions, or decide what is beautiful? Modern brain science, which began in the 1990s, has made considerable progress on this problem.[635]

A basic concept is *embodiment*, which in the dictionary is defined to be someone or something that is a perfect representative or example.[636]

Embodiment

Embodiment in brain science is best explained by an example. Suppose an adult shows a child how to measure the length of a wooden beam with a tape measure.

Thus the child learns the physical process of hooking the end of the tape to one end of the beam, extending the tape, and reading off the distance at the other end.

Brain science has shown that the brain stores these steps in specific locations that depend on the particular experience. The storage process is called *embodiment*.[637]

The brain later reruns the stored steps in the same region when the measuring of distance is to be repeated in some other setting. This includes muscle control of the hands, arms, and legs.

Amazingly, the stored steps are also rerun in that very same region when measuring length comes up in some other way—for example, when the sentence "Let's measure the length of this slat" is to be interpreted. In that case, the movement of legs, arms, and hands stored as part of the measuring process is not carried out. Regardless of the case, the rerunning of the steps is called *embodied simulation.*

Embodied Simulation

Embodied simulation has been proved not just for physical processes such as measuring length or opening a door with a key, but also for understanding pictures or movies, and for interpreting sound, smell, taste, and touch.

Embodied simulation is even used for the processing of abstract concepts such as hope or fear, and of emotions such as anger and sadness; and even for handling elusive concepts such as beauty.[638]

These results have been established in a multitude of studies,[639] and seeming conflicts of the concept of embodied simulation with other results have been suitably explained.[640]

Here is a second example of embodied simulation. Suppose an adult walks with a child along paintings in a museum and points out some paintings to be beautiful and others to be ugly.

The child stores the entire experience, including the paintings and the evaluation. Later, the same regions of the brain are used in embodied simulation to determine not only whether other paintings are beautiful or not, but also take part when the meaning of the sentence "This is a beautiful landscape painting" is to be determined.

It is astonishing that embodiment and the running of stored information by embodied simulation are universal: No matter what concepts are learned and stored in whatever regions of the brain, embodied simulation involving the same brain regions supports identification, classification, interpretation, or action in future events, each of which may be an actual occurrence, or a mentioning in a statement, or a depiction in a photo or movie, and so on. In summary:

"Every thought we have or can have, every goal we set, every decision or judgment we make, every idea we communicate makes use of the same embodied system we use to perceive, act, and feel. None of it is abstract in any way. Not moral systems. Not political ideologies. Not mathematics or scientific theories. And not language."[641]

Given the dominant role of embodied simulation in the interpretation of all aspects of the world, it is obvious that the brain can derive a reliable interpretation of a situation only if a sufficient number of relevant instances have been learned and can be evaluated with embodied simulation.

Put differently, if only a few, possibly nonrepresentative, cases are at hand, an interpretation based on those cases may well be deficient.

On the other hand, when a short phrase or keyword enters the brain and triggers embodied simulation of a large number of learned instances, the brain can produce a rich and nuanced interpretation of that phrase or keyword.

The six-word, likely apocryphal, short story by Hemingway cited in Chapter 12 triggers such impressive performance of the brain. The story is "Baby shoes. For sale. Never used."

Hearing the first sentence, the brain runs an embodied simulation of experiences with baby shoes, most if not all involving some joyful setting.

The sentence "For sale" triggers an embodied simulation of sales events listed in newspapers or posted on signs along a street, and so on.

Lastly, "Never used" characterizes a number of events where an item was acquired, never put to the intended use, and then sold since there was no future need.

The three results are assembled by the brain into a detailed story that depends on the experiences of the reader. For example, a baby did not survive birth and the previously bought shoes were no longer of use. Or it turned out that the feet of the newborn baby were already too large for the earlier purchased shoes. Thus, the interpretation may produce profound sadness or just a feeling of minor annoyance.

For a moment, let's go back to Chapter 12, where it was discussed whether sophisticated human communication requires recursion in human language. Embodiment and embodied simulation provide a strong argument that this is not so.

Indeed, due to those processes, rich images and relationships can be produced from the simplest sentences—for example, from Hemingway's six-word story. The same applies to the communication of the Pirahã people, where simple sentences without recursion produce vivid imagery of complex facts.[642]

The above discussion has left open how the brain uses the results of embodied simulation to produce coherent answers for questions or problems. For example, how does the brain tackle a philosophical question using embodied simulation?

As this book is written, in 2016, we do not know how this is done. There are some results for low-level processes of the brain that rely on embodied simulation,[643] but so far there are no results linking embodied simulation and high-level conclusions.[644]

But we dare to use the known facts about embodied simulation to explain why the brain may produce conflicting conclusions about philosophical problems.[645]

Explanation of Conflicting Conclusions

Once concepts are established by embodied simulation, the remaining steps of reasoning about philosophical problems are universally the same: With great care, the brain applies logic arguments to produce conclusions.[646] There may be flaws in those arguments, but they are quickly pointed out by others, and thus they are eliminated. Hence, we now conjecture the following:

Conclusions about philosophical problems essentially are produced by embodied simulation of prior experiences and application of logic.

The conjecture has the following corollary:

If the logic arguments made by several persons about a given philosophical problem are correct, then any conflicts of the resulting overall conclusions are due to embodiment of different learning experiences.

The above conjecture and corollary can be used to explain the contradictory conclusions of Chapter 8 produced by eminent mathematicians about the origin of mathematics: Since these persons surely did not use flawed logic arguments, the differences of the conclusions must be based on embodiment of different learning experiences. Let's look at three examples.

Gauss explored mathematics farther and wider than any contemporary. In addition, he carried out research in geodesy, geophysics, mechanics, electrostatics, astronomy, and optics.[647]

Accordingly, his philosophical statements were based on a broad range of experiences and, to this day, have rarely proved to be off the mark. In particular, his uncompromising vote for creation of mathematics, "[N]umber is purely a product of our mind."[648] should be seen in that light.

In contrast, Frege and Gödel focused in their work almost exclusively on the foundation of mathematics, particularly in logic, and acquired extensive and profound experience in that area.

Their votes for discovery reflect not only embodied simulation of that experience and their trust in logic arguments for philosophical questions, but also the frustration that a complete proof had eluded them.

Frege: "I hope I may claim in the present work to have made it probable that the laws of arithmetic are analytic judgments and consequently *a priori*."[649]

Gödel: "I am under the impression that after sufficient clarification of the concepts in question it will be possible to conduct these discussions with mathematical rigour and that the result will then be ... that the Platonistic view is the only one tenable."[650]

Similarly, one could try to characterize the experiences of each author of the other quotes of Chapter 8 about creation or discovery of mathematics, and thus motivate the diverse opinions.

How about today's opinions about the origin of mathematics?

Due to the victory of the axiomatic method in the first part of the 20th century,[651] the emphasis in mathematical developments has been on the formulation of axioms whenever a foundation is to be specified, thus moving the facts or circumstances motivating the work into the background.[652]

As a result, many mathematicians today learn to define a framework of axioms and then prove results starting from that foundation. Since the proof process is so neat and logical, it's easy to imagine that the derived results, and possibly the axioms, already exist somewhere—maybe in nature, or in a metaphysical world of ideas. Embodied simulation of that learning experience and logic reasoning then produce the conclusion that mathematics is discovered.[653]

Some mathematicians come to a different conclusion. They view mathematics as an art and a science, as stated by Borel in Chapter 8. Indeed, the work of mathematicians in *experimental mathematics*—an area spawned by ever faster computers—is best characterized as art and science. Here, computers are used to search for examples and counterexamples; find unusual patterns; produce proofs by exhaustive enumeration; validate or construct conjectures; prove equivalence of theories; evaluate infinite series, and find compact formulas for limit values; and investigate mathematical structures via visual inspection of computational results.[654]

The spirit of that work is similar to that of engineers and scientists, except that experimental mathematicians investigate mathematical objects,[655] while engineers and scientists explore, explain, and modify the world using mathematics as one of many tools.

As a result, the learning experiences of experimental mathematicians, engineers, and scientists differ greatly from those of mathematicians growing up with definitions and axioms. Embodied simulation then produces different conclusions about the origin of mathematics.[656]

We emphasize that the above statements about professions are offered with great humility. In a related quote, Gauss remarks on the afflictions of professions with a sense of humor:

"It may be true that men who are mere mathematicians have certain specific shortcomings, but that is not the fault of mathematics, for it is equally true of every other exclusive occupation. So there are *mere* philologists, *mere* jurists, *mere* soldiers, *mere* merchants, etc.

"To such idle talk it might further be added: that whenever a certain exclusive occupation is coupled with specific shortcomings, it is likewise almost certainly divorced from certain other shortcomings."[657]

As final topic of this chapter, let's use the above conjecture and corollary to understand why Wittgenstein's advice for the resolution of philosophical questions is so effective.

Wittgenstein's Advice

Chapter 9 introduces language games and demonstrates their use for the understanding of a given philosophical problem. It includes the claim that the operation of a language game produces *experiences about the world* in the brain that are relevant for understanding one facet; and that the experiences created by various language games empower the brain to resolve the dilemma.

We use the above conjecture about the role of embodied simulation to explain and justify that assertion.

When the brain operates language games related to a given philosophical question, it creates learning experiences that are stored in diverse regions of the brain. When that process stops, embodied simulation has a rich store of information available while processing the philosophical question.

The result is a nuanced evaluation that resolves the problem in the sense that it is declared to be nonsensical or, more simply, has been understood and does not require further investigation.

These conclusions apply only if the language games indeed produce experiences about the world. This happens when each language game is operated repeatedly and in great detail. In fact, any superficial operation would create just a temporary impression that is of little use for later evaluation of the given problem.

Chapter 9 discusses a number of philosophical questions and statements in the context of language games. In the next section, we look at one more case, the question "What is quality?"

The reader may want to not just read that material, but actually carry out the cited language games stated. Then it becomes evident that the games do create new experiences about the world that empower the brain for a more sophisticated evaluation of the question.

Example Question

Pervasive confusion about the question "What is quality?" is evident from the following assessment:

"In contemporary philosophy, the idea of qualities, and especially how to distinguish certain kinds of qualities from one another, remains controversial."[658]

The reader trying to think about the question will experience a strange, almost queasy feeling of confusion. As a remedy, the reader may want to go through instances where quality is declared, invoked, or dealt with.

Here are some examples.

- An engineer marvels at the accuracy of an astronomical clock.

- An engineer on an assembly line checks out features of the car being processed.

- A customer returns a sweater to the store, declaring that the zipper is not working.

- A guide in a museum declares a painting to be of high quality.

- A TV station announces that the air has a high pollen count.

- Parents spend quality time with their children.

Rasmus Sørnes's[659] astronomical Clock Number 4 is accurate within 7 seconds over 1,000 years. Completed in 1967.[660]

Piazza San Marco, Venice, by Giovanni Antonio Canal, better known as Canaletto, ca. 1730-1735, detail. Fogg Museum, Cambridge, Massachusetts. A superb painting of an architectural wonder.[661]

In the terminology of Wittgenstein, the reader is creating and operating a number of language games.

After a while, the reader has built up in the brain a large collection of experiences involving the word "quality" and is ready to think about the question "What is quality?"

When the reader now looks at the question about the nature of quality, embodied simulation will run the numerous, quite different instances where the word quality is used.

The conclusion will be that these results do not permit a statement about the nature of quality.

So the reader's ultimate response may be, "This is an ill-posed question that should be discarded as being inappropriate." Or possibly, "I now have a good understanding how the word 'quality' is used, and that's all there is to understanding the nature of that word."

Summary

Modern brain science has introduced the revolutionary concept of embodiment of experiences and subsequent evaluation by embodied simulation.

Based on that theory, we have posed a conjecture about the brain's processing of philosophical questions: The brain runs embodied simulation and evaluates the results using logic.

The conjecture helps explain why eminent mathematicians have come up with conflicting answers for the question "Is mathematics created or discovered?" Essentially, the differences are due to varied experiences inside and outside mathematics.

The conjecture also justifies the following recommendation for the resolution of any given philosophical problem: We should empower embodied simulation by studying numerous examples and language games before we try to address the problem.

Isn't it interesting that Wittgenstein suggested this very approach more than 70 years before the concept of embodied simulation was introduced in brain science?

———————————

The next, and final, chapter draws the conclusion about the origin of mathematics: It declares that creation takes place and not discovery.

14

Conclusion: Creation

The introductory chapter claims that mathematics is the most elaborate intellectual achievement of mankind.

The first half of this book, which covers the development of key parts of mathematics, shows that this claim is well justified. Indeed, it is safe to anticipate that most readers agree with that conclusion.

The introductory chapter also asserts that mankind created mathematics and didn't discover it.

The second half of the book demonstrates that this statement is correct. We use the wording "demonstrates" due to Wittgenstein's advice that philosophical conclusions can only be demonstrated and not proven. The demonstration consists of four parts.

First, Chapters 2–6 describe the struggle of mankind for mathematical concepts and results. Nowhere do we see that mathematicians were just discovering an existing result, say, in the way the continent Antarctica was discovered.

In fact, many times mathematical results turned out to be incomplete or erroneous, repeatedly requiring modification and correction.

Second, Chapter 11 examines the assertion that mathematics is unreasonably effective for modeling the world. If true, the claim supports the argument that mathematics is part of the natural world and thus is discovered. But detailed investigation reveals the claim to be incorrect.

Third, Chapter 12 looks at the assertion that mankind cannot function without mathematics. If true, then mathematics likely is part of nature and thus is discovered. But the example of an Amazonian tribe shows that humans can live without any mathematics whatsoever.

Fourth, in Chapter 10, a number of language games show that the assumption of discovery is quite untenable. We summarize the material of that chapter and the related discussion in Chapter 9.

The philosopher Wittgenstein proposes that we investigate philosophical questions or statements with language games. Indeed, that process generates insight for resolving the questions or statements.

We have constructed a number of language games for the question "Is mathematics created or discovered?"

Specifically, we observe how early man uses pebbles to represent sheep; we look over the shoulders of Bürgi as he reads Stifel's book on arithmetic; we listen to a fictitious discussion involving Riemann and Lebesgue; we sit in a class where a teacher guides the discussion of students about Euler's formula for polyhedra; we examine milestones of the development of music and compare them with similar ones in mathematics; and more.

Under the assumption of a discovered mathematics, the language games produce strange, sometimes bizarre consequences, thus discrediting that hypothesis.

Taken together, the four parts demonstrate that mathematics is created and not discovered.

That conclusion fits the historical facts about mathematical developments, does not require metaphysical concepts, and is consistent with the view that the human activities of music composition, sculpture, and writing are creation.

The conclusion also is consistent with the notion of Occam's razor[662] of science, according to which the simpler explanation is preferred when there is a choice.

There are now two possible outcomes:

The reader may agree with the above arguments and the main conclusion.

Or the reader may reject those arguments and simply insist on discovery of mathematics. For example, the reader may say, "Once the mathematical axioms are put in place, the rules of logic a priori determine what can ever be proved, and thus mathematics can only be discovered."

We don't know how to respond to such a statement. Of course, we could investigate it using additional language games, each of which would address a particular aspect.

For example, there would be language games concerning "The rules of logic a priori determine what can ever be proved." But then we anticipate that those language games would be rejected just like those laid out in detail in this book.[663]

Instead, we propose the following. If the reader does not believe that the above four parts demonstrate that mathematics is created, then we hope that the reader at least enjoyed reading about the wonderful and exciting history of mathematics and the relationships linking mathematics with other areas of human thought.

There is another way to look at the question of creation versus discovery, using the concept of a nonsensical statement proposed by Wittgenstein. Recall that the terms of such a statement collectively do not make sense and thus prohibit evaluation of the statement. An example nonsensical statement is "Rome is east of voltage." Wittgenstein argues that nonsensical statements are at the core of philosophical confusion.

How can we decide if a statement is nonsensical? In some cases it is easy, such as for the cited example: Rome and the direction east have no connection with the electricity concept voltage.

But in philosophical statements the nonsensical nature is usually hidden and sometimes so well camouflaged that detection of the defect is almost impossible.

For example, there is no simple analysis that reveals the question "What is quality?" to be nonsensical. No matter how carefully we apply logical arguments, resolution of the question somehow seems to escape us. But we can obtain that insight via language games.[664]

Let's apply this approach to the question at hand, or rather to a simpler statement for ease of discussion: "The human brain discovers mathematics." We want to show that this statement is nonsensical.

To this end, we investigate—with language games—how the human brain discovers results of any kind. We omit details and just mention that the games should include how scientists uncover facts about the universe, from geological features of the planets of the solar system down to properties of elementary particles; or how medical researchers unearth astonishing facts about organisms large and small.

We pause for a moment to clarify the discovery process of the cited examples. First, we *discover* some features of the world and represent them with initial models. In a trial and error process, we then expand the models to fit additional observations. So we *discover* facts about the world that so far have not been handled by our models or are even at odds with them. We then modify the models to accommodate the new facts. This process goes on and on, with no termination in sight. Thus, the features of the world are discovered, while the construction of models relies on human invention.

We compare the language games involving discovery with the development processes of mathematics as described in Chapters 2–6.

We see that the human development of mathematics is far different from human discovery processes—indeed is more like the creative activities of music composition, sculpture, and writing. As an aside, this also explains why mathematicians assess, and even debate, the *beauty* of a mathematical result or method.[665]

We conclude that we should not use the word "discovery" when referring to the development process of mathematics, in the same sense that we should not link a city, Rome, with a direction, east, and a concept of electricity, voltage. In other words, we conclude that the statement "The human brain discovers mathematics" is nonsensical.

Let's return to the question whether mathematics is created or discovered. We can reformulate the question by asking which of the statements "The human brain creates mathematics" and "The human brain discovers mathematics" is true.

We just have seen arguments that the latter statement is nonsensical, so we cannot determine if it is true. Indeed, we can only lay it aside and ignore it.

The statement "The human brain creates mathematics" is different. Indeed, Chapters 2–6 amply demonstrate that the development of mathematics is a creative process, and we accept the statement as true.

In Chapter 13 we conjecture that conclusions about philosophical problems essentially are produced by embodied simulation of prior experiences and application of logic.

We then propose that shortcomings of the first step—embodied simulation—should be remedied by the use of Wittgenstein's language games.

We venture to extend that notion to the following prediction: Brain science will produce scientific proof that the brain—so powerfully equipped by evolution to deal with the natural world—typically does not recognize a nonsensical philosophical statement as such.

That is, the brain is not equipped to detect the defective nature of a statement unless it obviously clashes with reality. The same applies to nonsensical questions. For example, "Rome is east of voltage" is immediately classified as nonsensical, while "What is quality?"[666] isn't easily recognized as falling into that category.

The predicted scientific insight into this shortcoming of the brain will have a significant effect on the discussion of philosophical statements: It will be deemed prudent to determine whether a given statement is nonsensical prior to any discussion whether the statement is true. We venture to predict that Wittgenstein's method of language games—so important for the arguments in this book— will play an important role in that process.

The conclusion of a created mathematics for us is most gratifying: It gives the human mind full credit for constructing the astonishingly large and rich edifice of mathematics. At last, this justifies the title of this book.

Notes

Throughout, "Wikipedia" refers to the English version unless another language is explicitly listed.

Chapter 1 Introduction

1. A Google search with "history of mathematics books" produces about 30 million results.

2. See Wikipedia "Florian Cajori" and "James Roy Newman." Several books by Cajori are available from archive.org, for example [Cajori, 1918], [Cajori, 1919a], [Cajori, 1919b], and [Cajori, 1928]. Also available is Newman's multi-volume *The World of Mathematics* [Newman, 1956].

3. See, for example, [Polkinghorne, 2011].

4. See Wikipedia "Occam's razor."

Chapter 2 Numbers

5. See Wikipedia "Ocean."

6. Source: https://en.wikipedia.org/wiki/Georg_Cantor#/media/File:Georg_Cantor2.jpg. "Georg Cantor2" by Unknown - http://i12bent.tumblr.com/post/3622180726/georg-cantor-german-mathematician-and-philosopher. Licensed under Public Domain via Commons.

7. [Rudman, 2007] covers the first 50,000 years of mathematics in clear and concise detail. The first part of this section relies on that material.

8. p. 54 [Rudman, 2007].

9. p. 68 [Rudman, 2007].

10. Proof: If there are only a finite number of primes, then multiply them all together and add 1. The resulting number cannot be divided by any of the primes in the collection, a contradiction of the claim that the given primes can produce all numbers.

11. Proof: First establish that, if a prime p divides $a \cdot b$, then p divides a or b.
Assume the opposite. Thus, for some integer m, $a \cdot b = p \cdot m$, and p does not divide a or b. Pick an instance with p as small as possible, and subject to that, a as small as possible. If $a > p$, the equation $(a - p) \cdot b = p(m - b)$ constitutes a smaller instance, a contradiction. If $a = p$ or $a = 1$, then p divides a or b, another contradiction. Thus, $1 < a < p$. Now a is the product of some prime, say q, and some $c \geq 1$. If q divides m, then $(a/q) \cdot b = p \cdot (m/q)$ is a smaller instance. This leaves the case where q does not divide m. Since q cannot divide the prime p, the equation $q \cdot (c \cdot b) = p \cdot m$ establishes a smaller instance, with q playing the role of p. Thus, all cases result in a contradiction, and the claim must hold.
We are ready to prove the unique factorization claim by induction. The result is immediate if the factored number is a prime. Otherwise, take any two factorizations. Select any prime factor of the first factorization. Split the second factorization arbitrarily into two parts. Then by the above result and induction, one of the two parts must contain the prime. Remove the prime from both factorizations, and repeat the argument. Thus, both factorizations are shown to be identical.

12. See Wikipedia "Fundamental Theorem of Arithmetic." The entry identifies the results by Euclid that directly imply the theorem.

13. [Sautoy, 2003].

14. Source: https://en.wikipedia.org/wiki/Euclid#/media/File:Euklid.jpg. "Euclid of Megara" (lat: Evklidi Megaren), Panel from the Series "Famous Men," Justus of Ghent, ca. 1474, Panel, 102cm x 80cm, Urbino, Galleria Nazionale delle Marche. This picture is meant to represent the famous mathematician Euclid of Alexandria, who in medieval times was wrongly identified with Euclid of Megara, the disciple of Socrates.

15. [Fritz, 1945].

16. $\sqrt{2}$ is the number that, when multiplied with itself, results in 2. In general, the kth root of a number n, denoted by $\sqrt[k]{n}$, is the number that, when multiplied together k times results in n.

17. Proof that $\sqrt{2}$ is irrational, using Euclid's result that each integer is a product of unique primes: Suppose $\sqrt{2}$ is rational, so for some integers a and b, $\sqrt{2} = \frac{a}{b}$, or equivalently $b \cdot \sqrt{2} = a$. Squaring each side, we get $b^2 \cdot 2 = a^2$. Now a and b are the unique product of some primes. Thus, by the uniqueness of the representations of a, a^2, b, and b^2, each prime factor of a^2 as well as of b^2 occurs an even number of times. Then the factorization of $b^2 \cdot 2$ has the prime 2 occurring an odd number of times, while it must occur in a^2 an even number of times, a contradiction. The proof is readily adapted to prove the following for any integer $k \geq 2$ and any positive integer n: If $\sqrt[k]{n}$ is not an integer, then it is not rational.

18. Source: https://www.math.ubc.ca/~cass/Euclid/ybc/ybc.html. Permission for use kindly granted by William A. Casselman, photographer and copyright holder. The clay tablet is part of the Yale Babylonian Collection. Provenance unknown, dated ca. 1800–1600 BCE. Purchased around 1912 by an agent of J. P. Morgan, who contributed it to Yale University as part of the foundation of its Babylonian Collection. For details about the numbers of the tablet and the computation of $\sqrt{2}$, see Wikipedia "Square root of 2."

19. Source: https://en.wikipedia.org/wiki/Richard_Dedekind#/media/File:Richard_Dedekind_1900s.jpg. "Richard Dedekind 1900s" by Unknown (Mondadori Publishers). Licensed under Public Domain via Commons.

20. The precise construction is as follows: Consider the rational numbers placed into a sorted sequence. To construct a single irrational number, cut the sorted sequence at some point, getting a sequence A below the cut and a sequence B above the cut. Select the cut in such a way that A has no greatest element. Then there is a unique real number that lies between A and B. If B has no least element, then the constructed number is irrational; otherwise it is rational. The figure below shows the construction of $\sqrt{2}$.
Consider the process repeated for all possible cuts. Then all real numbers are constructed.

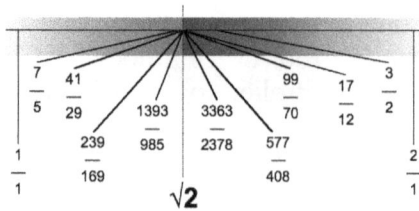

$\sqrt{2}$

(Source: https://commons.wikimedia.org/wiki/File:Dedekind_c
ut-_square_root_of_two.png#/media/File:Dedekind_cut-_squa
re_root_of_two.png. "Dedekind cut - square root of two" by Hy-
acinth - Own work. Licensed under Public Domain via Commons.)

21. More precisely, the polynomial has the form $a_n \cdot x^n + a_{n-1} \cdot x^{n-1} + \cdots + a_1 \cdot x^1 + a_0$, where n is some positive integer and all a_i are integers with $a_n \neq 0$.

22. The integers are algebraic since for any integer a, the polynomial $x - a$ has a as unique root. The rational numbers are algebraic numbers as well since, for integers a and b, $x = \frac{a}{b}$ is the unique root of the polynomial $b \cdot x - a$.

23. Source: https://commons.wikimedia.org/wiki/File:Leonhard
_Euler_2.jpg. "Leonhard Euler 2" by Jakob Emanuel Handmann -
2011-12-22 (upload, according to EXIF data). Licensed under Public
Domain via Commons.

24. Source: https://en.wikipedia.org/wiki/Joseph_Liouvil
le#/media/File:Joseph_liouville.jpeg. "Joseph liouville" by
http://www.math.sunysb.edu/. Licensed under Public Domain via
Commons.

25. [Erdös and Dudley, 1983].

26. See Wikipedia "Transcendental number."

27. The most famous equation constructed by Euler is $e^{i\pi} + 1 = 0$. The constant $e = 2.7182\ldots$ is now called the *Euler number*. The constant $\pi = 3.1415\ldots$ links diameter d and circumference c of a circle by $c = d \cdot \pi$. The physicist Richard Feynman called Euler's equation "our jewel" and "the most remarkable formula in mathematics;" see Wikipedia "Euler's formula."

28. See Wikipedia "Complex Plane."

29. Addition of complex numbers translates to addition of vectors, and multiplication is represented by a certain rotation of vectors.

See Wikipedia "Complex Number."

30. Source: https://en.wikipedia.org/wiki/Complex_number #/media/File:Complex_number_illustration.svg. "Complex number illustration" by Wolfkeeper. Licensed under CC BY-SA 3.0 via Commons.

31. See Wikipedia "e (mathematical constant)."

32. Source: https://en.wikipedia.org/wiki/Jacob_Bernoulli# /media/File:Jakob_Bernoulli.jpg. "Jakob Bernoulli" by Niklaus Bernoulli (1662-1716). Licensed under Public Domain via Commons.

33. Suppose an account holding $1 earns 100% interest in a year. If the interest is credited at the end of the year, the account will grow to $2. When the interest is credited twice a year at $\frac{100}{2} = 50\%$, the account will grow to $2.25. Now suppose the account is credited n times a year, with rate $\frac{100}{n}\%$. As we consider ever larger values of n, the account value at the end of the year approaches, and in the limit reaches, $e = 2.7182\ldots$.

34. See Wikipedia "e and pi are transcendental."

35. Source: https://en.wikipedia.org/wiki/Charles_Hermite# /media/File:Charles_Hermite_circa_1901_edit.jpg. "Charles_Hermite_circa_1901" by Unknown, derivative work: Quibik (talk) - Charles_Hermite_circa_1901.jpg. Licensed under Public Domain via Commons.

36. Source: https://en.wikipedia.org/wiki/Ferdinand_von_Lind emann#/media/File:Carl_Louis_Ferdinand_von_Lindemann.jpg. "Carl Louis Ferdinand von Lindemann" by Unknown - http://ww w.math.uha.fr/Pi/trans.html. Licensed under Public Domain via Commons.

37. [Rudman, 2007].

38. See Wikipedia "Leopold Kronecker."

39. See Wikipedia "Finitism."

40. Source: https://en.wikipedia.org/wiki/Leopold_Kronecker# /media/File:Leopold_Kronecker_1865.jpg. "Leopold Kronecker 1865" by Unknown. Licensed under Public Domain via Commons.

41. Source: https://en.wikipedia.org/wiki/Giuseppe_Peano#/

media/File:Giuseppe_Peano.jpg. "Giuseppe Peano" by Unknown - School of Mathematics and Statistics, University of St Andrews, Scotland [1]. Licensed under Public Domain via Commons.

42. See Wikipedia "Natural number."

43. There are a total of five axioms, listed below with comments in parentheses. The axioms use a constant symbol 0 and a unary function symbol S. ($S(n)$ is the successor $n + 1$ of a natural number n.)

1. 0 is a natural number. (The constant 0 is assumed to be a natural number.)
2. For every natural number n, $S(n)$ is a natural number.
3. For all natural numbers m and n, $m = n$ if and only if $S(m) = S(n)$. (S is an injection.)
4. For every natural number n, $S(n) = 0$ is false. (There is no natural number whose successor is 0.)
5. Let K be a set such that the following holds: 0 is in K, and for every natural number n, if n is in K, then $S(n)$ is in K. Then K contains every natural number. (Induction axiom.)

See Wikipedia "Giuseppe Peano" and "Peano Axioms" for further details.

44. See Wikipedia "Dedekind cut."

Chapter 3 Notation

45. Source: https://en.wikipedia.org/wiki/Differential_calculus#/media/File:Tangent_to_a_curve.svg. "Tangent to a curve" by Jacj. Later versions were uploaded by Oleg Alexandrov at en.wikipedia. - Transferred from en.wikipedia to Commons. Licensed under Public Domain via Commons.

46. Source: https://en.wikipedia.org/wiki/Integral#/media/File:Integral_example.svg. "Integral example" by I, KSmrq. Licensed under CC BY-SA 3.0 via Commons.

47. [Cajori, 1919b].

48. Source: https://en.wikipedia.org/wiki/Isaac_Newton#/media/File:GodfreyKneller-IsaacNewton-1689.jpg. "IsaacNewton-1689" by Godfrey Kneller, 1689. Licensed under Public Domain via Commons.

49. Source: https://en.wikipedia.org/wiki/Philosophi%C3%A6
_Naturalis_Principia_Mathematica#/media/File:Prinicipia
-title.png. "Principia-title" originally uploaded by Zhaladshar at
Wikisource - Transferred from en.wikisource to Commons.

50. [Cajori, 1919b] includes relevant excerpts of the Principia, including an English translation of the material taken from the third
edition.

51. Source: https://en.wikipedia.org/wiki/Gottfried_Wilhel
m_Leibniz#/media/File:Gottfried_Wilhelm_von_Leibniz.jpg.
"Gottfried Wilhelm von Leibniz" by Christoph Bernhard Francke -
/gbrown/philosophers/leibniz/BritannicaPages/Leibniz/Leibniz
Gif.html. Licensed under Public Domain via Commons.

52. Source: The Lilly Library, Indiana University, Bloomington, Indiana, kindly granted use of the photo.

53. For a detailed analysis, see pp. 85–95 [Grötschel et al., 2016].
Wikipedia "Leibniz–Newton calculus controversy" has a summary.

54. For example, the slope of the function f is denoted by $\dot{f}$, the
slope of the slope is $\ddot{f}$, and the integral is denoted by $\overline{f}$.

55. In modern mathematics, slope is denoted by $\frac{df}{dx}$, and the slope
of the slope by $\frac{d^2f}{dx^2}$. For the evaluation of the area under the function
$f(x)$ over the range $x = a$ to $x = b$, the expression is $\int_a^b f(x)dx$.

56. Source: https://en.wikipedia.org/wiki/Guillaume_de_l'H
ôpital#/media/File:Guillaume_de_l'Hôpital.jpg. "Guillaume
de
l'Hôpital." Licensed under Public Domain via Commons.

57. The title of the book is "Analyse des Infiniment Petits pour
l'Intelligence des Lignes Courbes." Recent research results confirm
that the book was largely based on the work of Johann Bernoulli. In
the introduction of the book, l'Hôpital acknowledges this, writing
"I must own myself very much obliged to the labours of Messieurs
Bernoulli, but particularly to those of the present Professor at Groe-
ningen [Johann Bernoulli], as having made free use of their Discoveries as well as those of Mr. Leibnitz: So that whatever they
please to claim as their own I frankly return them." The translation, slightly modified, is taken from http://faculty.wlc.edu/
buelow/calc/nt4-5.html. See also Wikipedia "Analyse des Infini-

ment Petits pour l'Intelligence des Lignes Courbes."

58. The proof that the slope of the function at a maximum or minimum must be 0, follows almost directly from the definition of slope: If at such a point the slope was positive, then a small increase of the variable would increase the function value, and a decrease of the variable would decrease the function value. If the slope was negative, the effects of the two changes of the variable would be reversed.

59. For details, see Wikipedia "Brachistochrone curve," including the mathematical description of the curve.

60. Source: https://en.wikipedia.org/wiki/File:Justus_Sus termans_-_Portrait_of_Galileo_Galilei,_1636.jpg. "Justus Sustermans - Portrait of Galileo Galilei, 1636" by Justus Sustermans. Licensed under Public Domain via Commons.

61. Source: https://en.wikipedia.org/wiki/Johann_Bernoulli #/media/File:Johann_Bernoulli2.jpg. "Johann Bernoulli2" by Johann Rudolf Huber. Licensed under Public Domain via Commons.

62. See Wikipedia "Brachistochrone curve."

63. See Wikipedia "Brachistochrone curve."

64. See Wikipedia "Calculus of variations."

65. Source: http://www.taralaya.org/science-park/BRACHIST OCHRONE_clip_image002.jpg. "Brachistochrone demonstration" photo copyright Jawaharlal Nehru Planetarium, Bangalore, India. The Planetarium kindly granted use of the photo.

66. Source: "Richard E. Bellman" copyright RAND Corporation, which kindly permitted use of the photo.

67. Proof that the segment New York – St. Louis is the shortest route: If that is not the case, then there is a shorter route connecting the two cities. But then that shorter route plus the St. Louis – Los Angeles segment constitute a shorter route New York – Los Angeles.

68. See Wikipedia "Dynamic Programming."

69. [Dreyfus, 1965].

70. Search Internet using "NASA dynamic programming" for a

long list of papers on control problems of space exploration that are solved with dynamic programming.

71. Chapter 4, 5 [Rudman, 2007].

72. The text was later translated into Arabic, Latin, and English. It became one of the most influential books in the history of mathematics; see [Heath, 1910].

73. Paragraph 101 [Cajori, 1928].

74. Source: https://en.wikipedia.org/wiki/Arithmetica#/media/File:Diophantus-cover.jpg. "Arithmetica by Diophantus." Licensed under Public Domain via Commons.

75. Folio 249 verso [Stifel, 1544]. See also Wikipedia "Michael Stifel."

76. Source: https://en.wikipedia.org/wiki/Michael_Stifel#/media/File:Michael_Stifel.jpeg. "Michael Stifel" by Unknown. Licensed under Public Domain via Commons.

77. For example, if we want to compute $\frac{1}{8} \cdot 4$, we look up the exponents of $\frac{1}{8}$ and 4, which are -3 and 2, add them together and get -1, and finally look up for the latter exponent the corresponding number, which is $\frac{1}{2}$. Thus, $\frac{1}{8} \cdot 4 = \frac{1}{2}$.

78. For example, to find the cube of a number, its exponent is multiplied by 3; to take the cubic root, the exponent is divided by 3.

79. See de.wikipedia "Jost Bürgi."

80. Source: https://en.wikipedia.org/wiki/Jost_B%C3%BCrgi#/media/File:Jost_B%C3%BCrgi_Portr%C3%A4t.jpg. "Jost Bürgi Porträt" by User Dvoigt on de.wikipedia. Licensed under Public Domain via Commons.

81. Source: https://en.wikipedia.org/wiki/Jost_B%C3%BCrgi#/media/File:JostBurgi-MechanisedCelestialGlobe1594.jpg. "Jost Buergi-Mechanised Celestial Glob2" photo by Horology - Own work. Licensed under CC BY-SA 3.0 via Commons.

82. Bürgi had an exponent notation for variables that is connected with Stifel's exponent. He wrote a Roman numeral above a constant as the exponent of a variable; the constant was the coefficient of the variable. For example, the term $\overset{vi}{4}$ has the Roman numeral $vi = 6$ above the number 4; the expression denotes $4x^6$ in modern

notation; see paragraph 296 [Cajori, 1928].

83. Bürgi's choice of $B = 1.0001$ was really clever since successive multiplication by B amounted to the following: Take the number already on hand for a given n value, write it again below while shifting it four positions to the right, and add the two numbers. Voilà, there is the number for $n + 1$. Thus, the numbers for the exponents $n = 0, \ldots, 23027$ could readily be computed, starting with $n = 0$.

Some additional steps were needed to avoid build-up of numerical errors; see p. 15 [Waldvogel, 2012]. But even with such checks and corrections, the entire construction process of the tables was extraordinarily efficient. The estimate of several months of manual effort is included on p. 15 [Waldvogel, 2012].

84. p. 6 [Waldvogel, 2012].

85. [Bürgi, 1620].

86. The manual is available on pp. 26–36 [Gieswald, 1856].

87. Source: Toggenburger Museum, Lichtensteig, Switzerland. The museum kindly has granted permission to use the photo.

88. See Wikipedia "John Napier."

89. p. 188, 189 [Cajori, 1919a] explains Napier's line of thought as follows. Imagine two lines, the first one with two marks A and B, and second one with just one mark D. A point moves from A to B, and a second point from D in one of the two possible directions. The points start at the same time and with the same initial velocity. The first point immediately starts to slow down: At any intermediate point between A and B, say C, the initial velocity is reduced by the factor (distance C to B)/(distance A to B). In contrast, the second point never changes its velocity.

Suppose the first point has arrived at C as just stated, and the second one is at a point F at the same time. Napier calls the distance from D to F the *logarithm* of the distance from C to B.

The modern-day logarithm function is 0 when its argument is 1; in Napier's tables, the distance from A to B is 10^7, so 0 is Napier's logarithm of 10^7. Thus, his approach didn't suggest the concept of *base*; indeed, that concept seemingly is inapplicable to his method. But when movements of the two points are analyzed, it becomes clear that Napier's logarithm for a number x is equal to $10^7 (\ln 10^7 - \ln x)$, where $\ln$ denotes the *natural logarithm*, which has as base the

Euler number $e = 2.7182\ldots$

90. Source: https://en.wikipedia.org/wiki/John_Napier#/med
ia/File:John_Napier.jpg. "John Napier" by Unknown - scanned
from http://www-history.mcs.st-and.ac.uk/history/PictDispl
ay/Napier.html. Licensed under Public Domain via Commons.

91. Source: https://en.wikipedia.org/wiki/John_Napier#/media
/File:Logarithms_book_Napier.jpg. "Logarithms book Napier"
by Unknown - Napier, Mark (1834), William Blackwood. Licensed
under Public Domain via Commons.

92. See Wikipedia "Henry Briggs (mathematician)."

93. [Gieswald, 1856] contains a detailed discussion. For additional
information, see [Waldvogel, 2012] and Wikipedia "History of logarithms."

94. See Wikipedia "René Descartes."

95. [Descartes, 1637].

96. Source: https://en.wikipedia.org/wiki/Ren%C3%A9_Desc
artes#/media/File:Frans_Hals_-_Portret_van_Ren%C3%A9_D
escartes.jpg. "Frans Hals - Portret van René Descartes" after
Frans Hals (1582/1583–1666) - André Hatala [e.a.] (1997) De eeuw
van Rembrandt, Bruxelles: Cédit communal de Belgique, ISBN 2-
908388-32-4. Licensed under Public Domain via Commons.

97. Source: https://en.wikipedia.org/wiki/Discourse_on_the
_Method#/media/File:Descartes_Discours_de_la_Methode.jpg.
"Discours de la Méthode" by Unknown. Licensed under Public Do-
main via Commons.

98. [Rudman, 2007].

99. Archimedes by Domenico Fetti, 1620. Source: https://en.wik
ipedia.org/wiki/File:Domenico-Fetti_Archimedes_1620.jpg.
Painting of Alte Meister Museum, Dresden, Germany; see http:
//archimedes2.mpiwg-berlin.mpg.de/archimedes_templates/p
opup.htm. Public Domain under US copyright code PD-old-100.

100. "Archimedes parabola with triangle" by K. Truemper, released
into Public Domain under Creative Commons CC0.

101. Source: https://en.wikipedia.org/wiki/Archimedes#/me
dia/File:Esfera_Arqu%C3%ADmedes.jpg. "Esfera Arquímedes"

by Andertxuman - Own work. Licensed under Public Domain via Commons.

102. [Netz and Noel, 2007] contains historical and mathematical details; see also Wikipedia "Archimedes."

103. Paragraph 340 [Cajori, 1928].

104. Paragraph 298 [Cajori, 1928].

105. See Wikipedia *"La Géométrie"* and "Cartesian coordinate system."

106. Source: `https://en.wikipedia.org/wiki/Cartesian_coordi nate_system#/media/File:Cartesian-coordinate-system.svg`. "Cartesian coordinate system." By K. Bolino - Made by K. Bolino (Kbolino), based upon earlier versions. Licensed under Public Domain via Commons.

107. Source: `https://en.wikipedia.org/wiki/Cartesian_coordi nate_system#/media/File:Cartesian-coordinate-system-with -circle.svg`. "Cartesian-coordinate-system-with-circle" by 345Kai. Licensed under CC BY-SA 3.0 via Commons.

108. See Wikipedia "History of the function concept."

109. Source: `https://en.wikipedia.org/wiki/Function_(ma thematics)#/media/File:Function_machine2.svg`. "Function machine2" by Wvbailey (talk) - Own work (Original text: I created this work entirely by myself.). Licensed under Public Domain via Commons.

110. Source: `https://en.wikipedia.org/wiki/Graph_of_a_funct ion#/media/File:Three-dimensional_graph.png`. "Three-dimensional graph" by dino (talk) - Own work. Licensed under CC BY-SA 3.0 via Commons. The depicted function is $f(x,y) = sin(x^2) \cdot cos(y^2)$.

111. A function is continuous if changes of $f(x)$ can always be made arbitrarily small by suitably restricting changes of x.

112. See Wikipedia "Alan Turing" and "Halting problem."

113. Source: `https://en.wikipedia.org/wiki/Alan_Turing#/m edia/File:Alan_Turing_Aged_16.jpg` "Alan Turing Aged 16" by Unknown - `http://www.turingarchive.org/viewer/?id=521&tit le=4`. Licensed under Public Domain via Commons.

114. The natural number representing a computer program is determined as follows. Each instruction of the computer program is coded by a natural number. The string of these natural numbers, viewed as one large natural number, then represents the entire computer program.

115. The results claimed for the number of functions follow directly from Cantor's work discussed in Chapter 4. Essentially, the number of functions with the natural numbers as input and 0 or 1 as output is the same as the number of real numbers, and the number of functions with the real numbers as input and 0 or 1 as output is equal to the number of subsets of the set of real numbers. In the notation of Chapter 4, the first number is $2^{\aleph_0}$, which is infinitely larger than the number of natural numbers $\aleph_0$. The second number is $2^{2^{\aleph_0}}$, which is infinitely larger than $2^{\aleph_0}$.

116. See Wikipedia "Function (mathematics)."

117. Source: https://en.wikipedia.org/wiki/Bernhard_Riemann #/media/File:Georg_Friedrich_Bernhard_Riemann.jpeg. "Georg Friedrich Bernhard Riemann" by https://de.wikipedia.org/wik i/Bernhard_Riemann#/media/File:BernhardRiemannAWeger.jpg of de.wikipedia. Licensed under Public Domain via Commons.

118. Source: https://commons.m.wikimedia.org/w/index.php?se arch=Henri+Lebesgue+&fulltext=search#/media/File%3ALebesgu eH.gif. By Unknown. Public Domain since copyright has expired.

119. In mathematical notation, the width of a slice is dx, the height of a slice is $f(x)$, and the estimate of the area of the slice is $f(x)dx$. When dx becomes ever smaller, then under suitable assumptions the sum of these areas, denoted by $\int f(x)dx$, converges to the desired value.

120. For a vertical strip of width dx, estimation of the strip's area by $f(x)dx$ is successful only if the function does not jump within the interval. Hence, any function that jumps within the interval, no matter how narrow, cannot be processed.

121. Source: https://en.wikipedia.org/wiki/Lebesgue_int egration#/media/File:RandLintegrals.png. "RandLintegrals" included in en.wikipedia. Licensed under CC BY-SA 3.0 via Commons.

122. In mathematical notation, the height of a horizontal slice is dt,

and the length of the slice is the sum of the lengths of the segments of the x-axis for which the function value is greater than t. Using a function μ that can add up the length of these line segments, the length of the slice is computed as $f^*(t) = \mu(\{x \mid f(x) > t\})$. The Lebesgue integral of f is then $\int f \, d\mu = \int_0^\infty f^*(t) \, dt$ where the integral on the right is a Riemann integral.

123. See Wikipedia "Lebesgue integration."

124. See Wikipedia "Lebesgue integration."

125. The Lebesgue integral can be computed since the total length of the line segments for which $f(x) > 0$ can be proved to be 0.

Chapter 4 Infinity

126. Leonardo Donato rebelled against the power of Pope Paul V in a struggle whose outcome is best characterized as a draw; see Wikipedia "Leonardo Donato."

127. Source: https://commons.wikimedia.org/wiki/File:Gali leo_Donato.jpg#/media/File:Galileo_Donato.jpg. "Galileo Donato" by H. J. Detouche - http://www.astro.unipd.it/insap6/mainPage.html. Licensed under Public Domain via Commons.

128. Source: https://en.wikipedia.org/wiki/Nicolaus_Cop ernicus#/media/File:Nikolaus_Kopernikus.jpg. "Nikolaus Kopernikus" by Unknown - http://www.frombork.art.pl/An g10.htm. Licensed under Public Domain via Commons.

129. pp. 83–85 [Alexander, 2014].

130. p. 138 [Alexander, 2014]. See also Wikipedia "Galileo Galilei."

131. [Alexander, 2014] describes in detail the new ideas and the forces trying to suppress them.

132. See Wikipedia "Bonaventura Cavalieri."

133. Source: https://en.wikipedia.org/wiki/Bonaventura_Cava lieri#/media/File:Bonaventura_Cavalieri.jpeg. "Bonaventura Cavalieri." Licensed under Public Domain via Common.

134. Source: https://archive.org/. Search for "Bonaventura Cavalieri." Public Domain .

135. See Wikipedia "Cavalieri's principle."

136. *Geometria indivisibilibus continuorum nova quadam ratione* is available at https://archive.org/. Search for "Bonaventura Cavalieri."

137. See Wikipedia "Bonaventura Cavalieri."

138. See Chapter 3.

139. See Wikipedia "Evangelista Torricelli."

140. Source: https://en.wikipedia.org/wiki/Evangelista_Torr
icelli#/media/File:Evangelista_Torricelli_by_Lorenzo_Li
ppi_%28circa_1647,_Galleria_Silvano_Lodi_%26_Due%29.jpg.
"Evangelista Torricelli" by Lorenzo Lippi, ca. 1647. Galleria Silvano
Lodi & Due. Licensed under Public Domain via Commons.

141. Source: Joern Koblitz of Milestones of Science Books, Bremen,
Germany, kindly granted permission to use photo of title page of
Torricelli's "Opera Geometrica," all rights reserved.

142. Source: "Rectangle and Parallelogram" by K. Truemper, released into Public Domain under Creative Commons CC0.

143. Source: "Hyperbola" by K. Truemper, released into Public Domain under Creative Commons CC0.

144. Source: https://en.wikipedia.org/wiki/Gabriel%27s_Horn
#/media/File:GabrielHorn.png. "Gabriel's Horn" by RokerHRO -
Own work. Licensed under Public Domain via Commons.

145. See Wikipedia "Gabriel's Horn" for details of the following
formulas. The volume V and surface area A of the trumpet trun-
cated at some point $a > 1$ are given by $V = \pi \int_1^a \left(\frac{1}{x}\right)^2 dx =$
$\pi \left(1 - \frac{1}{a}\right)$ and $A = 2\pi \int_1^a \frac{1}{x}\sqrt{1 + \left(\frac{-1}{x^2}\right)^2} dx > 2\pi \int_1^a \frac{1}{x} dx = 2\pi \ln a$.
When a goes to ∞, we get $\lim_{a \to \infty} V =$
$\lim_{a \to \infty} \pi \left(1 - \frac{1}{a}\right) = \pi$ and $\lim_{a \to \infty} A \geq \lim_{a \to \infty} 2\pi \ln a = \infty$.

146. It's easy for us today to exhibit the flaw in the paint argument.
At the time, the finite volume and infinite surface of the trumpet
seemingly were a paradox, and Torricelli attempted several proofs
to show that the surface was finite. See Wikipedia "Evangelista Torricelli."

147. See Chapters 3–5 of [Alexander, 2014] for details about the
fight of the Jesuits against the idea of indivisibles and the devastating impact of their victory.

148. Source: `https://en.wikipedia.org/wiki/John_Wallis#/med ia/File:John_Wallis_by_Sir_Godfrey_Kneller,_Bt.jpg`. Public Domain under US copyright code PD-old-100 .

149. Source: go to `https://archive.org/` and search for "john wallis de sectionibus conicis." Licensed under Public Domain Mark 1.0 Creative Commons.

150. See Wikipedia "John Wallis."

151. Here is an example of Wallis's method. In Proposition 3 of *De sectionibus conicis*, he computes the area of triangular figures as follows. He first slices the area of such a figure horizontally into an infinite number of strips. Let the base line of the figure have length B and the height of the figure be A. Since the number of strips is equal to ∞, each strip has height $\frac{A}{\infty}$. Since the length of the strips goes from B to 0 evenly as one moves from the baseline to the top point of the figure, the average strip length is $\frac{B}{2}$. Since there are an infinite number of strips, the total length of all strips is ∞ times the average strip length, that is, $\infty \cdot \frac{B}{2}$. The total area is the product of strip height times total strip length, so Area $= \frac{A}{\infty} \cdot \infty \cdot \frac{B}{2}$. The two instances of ∞ cancel out, and thus Area $= \frac{A \cdot B}{2}$.

152. See Wikipedia "John Wallis."

153. See Wikipedia "Wallis product."

154. At the time of Newton and Leibniz, the function concept was yet to be introduced; see Chapter 3. Instead, Newton and Leibniz treated formulas. For clarity, we use the function concept.

155. See Chapter 3.

156. We use d for a small change of x to discuss both Newton's and Leibniz's methods. Newton actually employed o for a small quantity, while Leibniz used dx.

157. Source: "Differential" by K. Truemper, released into Public Domain under Creative Commons CC0.

158. The description of Leibniz's process in terms of x, $f(x)$ and d is necessarily imprecise since the function concept was unknown at that time. We use discussion of the $f(x) = x^2$ case by Bernoulli as cited on p. 28 [Bos, 1974], which roughly matches the description used here.

159. p. 9 [Cajori, 1919b] contains Newton's arguments:
"It is objected, that there is no ultimate proportion of evanescent quantities [here, $f(x+d) - f(x)$ and d are the evanescent quantities]; because the proportion, before the quantities have vanished, is not ultimate; and when they have vanished, is none. But, by the same argument, it might as well be maintained, that there is no ultimate velocity of a body arriving at a certain place, when its motion is ended: because the velocity, before the body arrives at the place, is not its ultimate velocity; when it has arrived, is none.
"But the answer is easy: for the ultimate velocity is meant that, with which the body is moved, neither before it arrives at its last place, when the motion ceases, nor after; but at the very instant when it arrives; that is, the very velocity with which the body arrives at its last place, when the motion ceases.
"And, in like manner, by the ultimate ratio of evanescent quantities is to be understood the ratio of the quantities, not before they vanish, nor after, but that with which they vanish."

160. See the overview of Leibniz's method in [Bos, 1974].

161. Source: https://en.wikipedia.org/wiki/Bernard_Bolzano#/media/File:Bernard_Bolzano.jpg. "Bernard Bolzano." Licensed under Public Domain via Commons.

162. Source: https://en.wikipedia.org/wiki/Augustin-Louis_Cauchy#/media/File:Augustin-Louis_Cauchy_1901.jpg. "Augustin-Louis Cauchy 1901" by Library of Congress Prints and Photographs Division. From an illustration in: Das neunzehnte Jahrhundert in Bildnissen / Karl Werckmeister, ed. Berlin: Kunstverlag der Photographischen Gesellschaft, 1901, vol. V, no. 581. Licensed under Public Domain via Commons.

163. The technical definition of convergence of a sequence $S = s_1, s_2, s_3, \ldots$ to a limit L is as follows: For any value $\delta > 0$, there is an index n such that, for all $i > n$, the terms s_i satisfy $|s_i - L| < \delta$.

164. The stated definition of continuity isn't practically useful: To verify continuity just at one point c, we must consider all sequences $x_1, x_2, x_3, \ldots$ that converge to c, a potentially difficult task.
Weierstrass created an alternate and eminently useful definition using ϵ and δ: For any tolerance $\epsilon > 0$—no matter how small—there is a value $\delta > 0$ such that, whenever x satisfies $c - \delta < x < c + \delta$, then $f(c) - \epsilon < f(x) < f(c) + \epsilon$.
The condition can be stated in terms of the drawing below as fol-

lows: Given any $\epsilon > 0$—no matter how small—let the horizontal borders of the horizontal strip be defined by $f(c) - \epsilon$ and $f(c) + \epsilon$ as shown. Then there must be $\delta > 0$ defining the vertical borders of the vertical strip by $c - \delta$ and $c + \delta$, such that the portion of the function $f(x)$ falling within the vertical strip, here indicated by a heavy segment, is also contained in the horizontal strip.

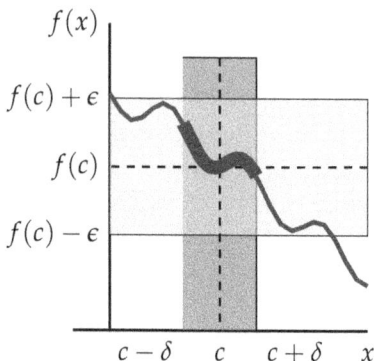

(Source: "Delta/Epsilon Continuity Definition" by K. Truemper, released into Public Domain under Creative Commons CC0.)

165. Source: `https://en.wikipedia.org/wiki/Karl_Weierstrass#/media/File:Karl_Weierstrass.jpg`. "Karl Weierstrass." Licensed under Public Domain via Commons.

166. Source: "Continuous function" by K. Truemper, released into Public Domain under Creative Commons CC0.

167. Source: "Function with jump" by K. Truemper, released into Public Domain under Creative Commons CC0.

168. See Wikipedia "Limit (mathematics)."

169. Source: "Extended Function" by K. Truemper, released into Public Domain under Creative Commons CC0.

170. See Wikipedia "Mathematical analysis."

171. For discussion of the case of a general function $f(x)$, let's use Leibniz's dx instead of d and the limit notation. We then have the derivative $\frac{df}{dx} = \lim_{dx \to 0} \frac{f(x+dx)-f(x)}{dx}$. The limit notation emphasizes that the slope of $f(x)$ at a point x is computed by finding a continuous extension of the function $\frac{f(x+dx)-f(x)}{dx}$ where x is an arbitrary point, dx is the variable, and the extension is to be found for $dx = 0$. How do we know that this extension will always exist?

Well, we do not. For example, if the function jumps at a point x, then $f(x + dx) - f(x)$ jumps at $dx = 0$, and $\frac{f(x+dx)-f(x))}{dx}$ has no continuous extension at $dx = 0$.

172. For a long time, it was believed that slope could always be computed for continuous functions except for some exceptional points. For the Weierstrass function, the desired continuous extension of $\frac{f(x+dx)-f(x)}{dx}$ does not exist for *any* point x. In fact, it is now known that virtually *all* continuous functions exhibit this behavior! See Wikipedia "Weierstrass function" for details.
As an aside, suppose we declare that the continuous functions employed in the sciences and engineering, which typically are differentiable except maybe at some special points, are *normal*, and that nondifferentiable continuous functions like the Weierstrass function are *pathological* cases. Then according to the cited result, virtually all continuous functions are pathological, a strange conclusion. It is one more indication that mathematics is not part of nature.

173. See Wikipedia "Finitism."

174. See Wikipedia "Actual infinity."

175. Cantor compares the set N of natural numbers with any other infinite set T as follows.

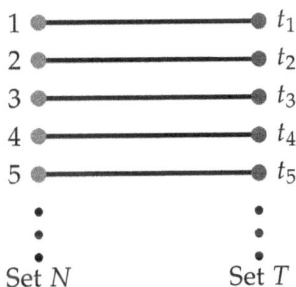

Set N Set T

(Source: "Countable set" by K. Truemper, released into Public Domain under Creative Commons CC0.)
First, he checks if each element of N can be assigned to a different element of T as shown above, where the elements $1, 2, 3, \ldots$ on the left are in N and the elements $t_1, t_2, t_3, \ldots$ on the right in T. When this is the case, he declares that the cardinality of T is greater than or equal to that of N. In shorthand notation, $|T| \geq |N|$.
Then he checks if each element of T can be assigned to a different element of N. When that is the case, he has $|T| \leq |N|$. When both cases apply, he declares that N and T have the same cardinality, denoted by $|N| = |T|$. The same approach is used to compare the

cardinality of arbitrary infinite sets.

176. It is by no means self-evident that the set N of natural numbers has the smallest cardinality of all infinite sets. But when the *axiom of countable choice* is added to Zermelo-Fraenkel set theory—more on this in Chapter 6—then it can be proved that N indeed represents the smallest infinity. Indeed, it can be shown that any infinite set T contains a countable infinite subset, that is, a subset of the form $\{t_1, t_2, t_3, \dots\}$. Hence the elements of N can be assigned to the elements of the subset, and $|N| \leq |T|$ has been proved. For details, see Wikipedia "Axiom of countable choice."

177. Proof: Let $R = \{r_1, r_2, r_3 \dots\}$ and $S = \{s_1, s_2, s_3 \dots\}$. Create T from R by replacing each element r_i by the elements of S, notationally modified so that they are unique, say denoted by $s_1^i, s_2^i, s_3^i, \dots$. There are just as many s_j^i in T as there are pairs (r_i, s_j). For simplicity, assume that each r_i and s_j is a natural number. By the fundamental theorem of arithmetic—see Chapter 2—the prime numbers 2 and 3 can be used to uniquely represent each pair (r_i, s_j) by the natural number $2^{r_i} \cdot 3^{s_j}$. The set of the latter numbers is a subset of the natural numbers and thus countable, and so is T.

178. Each rational number is a ratio of two integers m and n. Thus, by substitution, the ratios can be counted. For the algebraic numbers, the polynomials with integer coefficients can be counted, and each of them has a finite number of solutions. By substitution, the solutions can be counted.

179. Proof: Use the fact that the set of algebraic numbers is countable, then use repeated substitution to handle n-dimensional space.

180. Consider the real numbers between 0 and 1, represented in binary notation. An example is $0.1001011\dots = 2^{-1} + 2^{-4} + 2^{-6} + 2^{-7}\dots = 0.58\dots$ in decimal notation. It suffices to show that there are more real numbers than natural ones, since it has already been shown that the number of natural numbers matches that of the rational and algebraic numbers.
The result is established by showing that, when one tries to assign the real numbers to the natural numbers, then not all of them can be accommodated. Assume such an assignment is possible, say where for $i = 1, 2, 3, \dots$, the real number r_i is assigned to i. Since all digits of each r_i are either 0 or 1, define a new number r^* whose ith digit is the opposite of the ith digit of r_i. Thus, r^* is different from all r_i and thus has not been assigned, a contradiction. For further details

about the diagonal argument, see Wikipedia "Cantor's diagonal argument."

181. Consider the binary representation of the real numbers r between 0 and 1. Take the example $r = 0.1001011\ldots$, and look at the 1s of r. They are in positions 1, 4, 6, 7, ... These indices correspond to the subset $\{1, 4, 6, 7, \ldots\}$ of the set N natural numbers, so the subset is a unique representation of r. Conversely, the subset uniquely defines the 1s of r, and r is a unique representation of the subset. Thus, the subsets of N are in one-to-one correspondence with the real numbers r between 0 and 1.

The subsets of a given set S form a set called the *power set* of S. If S is finite and has n elements, then the power set of S has 2^n elements.

Cantor extends this notation to infinite sets. Since N has cardinality $\aleph_0$, he declares that the power set of N has cardinality $2^{\aleph_0}$. Due to the above derived correspondence of the power set of N and the set of real numbers between 0 and 1, the latter set has cardinality $2^{\aleph_0}$ as well. By a trivial substitution step, the same conclusion applies to the set of all real numbers.

182. The expanded conclusion for the algebraic numbers relies on the fact that the proof of the original statement involving the rational numbers uses only the fact these numbers are countable. Now the algebraic numbers are countable as well, and the result follows.

183. The natural numbers allow an ordering of finite sets. For example, suppose we have a set with six elements. We can label them $a_1, a_2, \ldots, a_6$ and thus order them.

Cantor's concept of *ordinal numbers*, for short *ordinals*, extends this idea to infinite sets. Given an infinite set, ordinal numbers can be used to label the elements and thus obtain an ordering.

Ordinals are different from the *cardinal numbers*, for short *cardinals*, such as $\aleph_0$: The latter concept just measures the number of elements in a set and is not concerned with any ordering of the elements. Cantor introduced ordinals so that he could define and manipulate infinite sequences.

The finite ordinals, as well as the finite cardinals, are just the natural numbers, including the 0. So we have 0, 1, 2, ... The smallest infinite ordinal is labeled ω and is associated with the cardinal $\aleph_0$. But for the many possible sets with cardinality $\aleph_0$, a huge number of orderings are possible. This is reflected in a correspondingly vast number of ordinals. An appealing intuitive definition of these and subsequent ordinals is given in Wikipedia "Ordinal number," as follows.

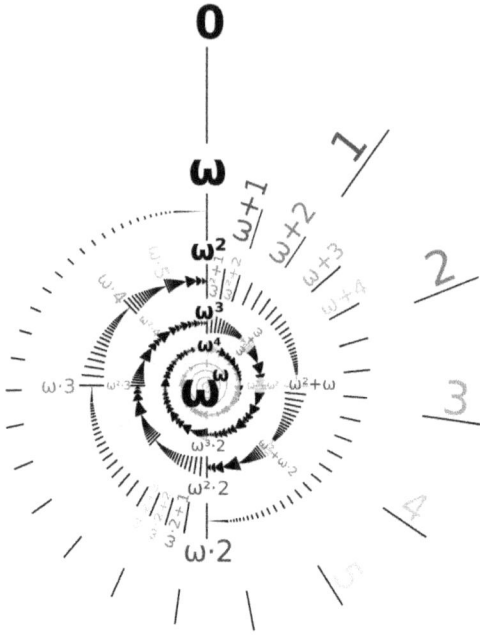

Start with the natural number 0 of the above spiral, and proceed clockwise. The next ordinal numbers are 1, 2, ... After all natural numbers comes the first infinite ordinal, ω, and after that $\omega + 1$, $\omega + 2$, $\omega + 3$, ... Ignore the meaning of the "+" sign, and consider these terms just to be labels.

Next comes $\omega + \omega$, denoted by $\omega \cdot 2$, then $\omega \cdot 2 + 1$, $\omega \cdot 2 + 2$, ..., then $\omega \cdot 3$, and later $\omega \cdot 4$. Now the ordinals $\omega \cdot m + n$, where m and n are natural numbers, formed by this process are followed by ω^2. Going on, we encounter ω^3, ω^4, ..., then ω^ω.

At that point, the spiral stops, but the list of ordinals actually goes on. Next are ω^{ω^2}, ω^{ω^3}, ..., and much later an ordinal called ϵ^0. And this constitutes a list of relatively small, countable ordinals! Indeed, we can continue this construction indefinitely. The smallest uncountable ordinal is the set of all countable ordinals, expressed as ω_1. Cantor denotes the cardinality of that set by $\aleph_1$.

184. See Wikipedia "Georg Cantor."

185. Source: https://en.wikipedia.org/wiki/G%C3%B6sta_Mitta
g-Leffler#/media/File:Magnus_Goesta_Mittag-Leffler_1.jpg.
"Magnus Goesta Mittag-Leffler 1" by Unknown. Licensed under
Public Domain via Commons.

186. The letters are summarized in [Schoenflies, 1927], written for
the celebration of Mittag-Leffler's 80th birthday. The paper brings
out Mittag-Leffler's central role in the support of Cantor's work.
As an aside, Mittag-Leffler was a strong advocate for women. For
example, because of his efforts, Sofia Kovalevskaya, the first major
Russian female mathematician, became the first woman anywhere
in world to hold the position of full professor at a university; see
Wikipedia "Mittag-Leffler" and "Sofia Kovalevskaya."

Chapter 5 Six Problems of Antiquity

187. See Chapter 3.

188. The latter effort culminated in Euler's famous equation $e^{i\pi} +
1 = 0$, which brings together the transcendental e and π, the imag-
inary i, and the fundamental 0 and 1. The physicist Richard Feynman called the equation "our jewel" and "the most remarkable formula in mathematics." See Wikipedia "Euler's formula."

189. See Chapter 4.

190. See Wikipedia "Compass equivalence theorem" for the simple
construction.

191. Source: "Non-collapsing compass" by K. Truemper, released
into Public Domain under Creative Commons CC0.

192. "Triangle, square, and pentagon" by K. Truemper, released
into Public Domain under Creative Commons CC0.

193. Source: "Parabola" by K. Truemper, released into Public Domain under Creative Commons CC0.

194. Source: "Trisection" by K. Truemper, released into Public Domain under Creative Commons CC0.

195. Source: "Double cube" by K. Truemper, released into Public
Domain under Creative Commons CC0.

196. Source: "Square circle" by K. Truemper, released into Public
Domain under Creative Commons CC0.

197. There are a number of equivalent formulations of the fifth axiom; see Wikipedia "Euclidean geometry." We list Playfair's version; see Wikipedia "Playfair's axiom." Note that the first four axioms imply that there is at least one parallel line. Thus, the condition of Playfair's axiom that there is *at most* one parallel line means that there is *exactly* one such line.

198. Source: "Parallel line" by K. Truemper, released into Public Domain under Creative Commons CC0.

199. Source: Archiv of the Berlin-Brandenburg Academy of Sciences (ABBAW), Department Collection, Portraits of Scientists, C. F. Gauss, ZIMM-0001. Photo by Stephan Fölske. The Academy has kindly granted permission for use of the photo.
The painting has a complex history: In 1840, the Dutch painter Christian Albrecht Jensen created the Gauss portrait for the astronomical observatory in Pulkovo, a village near St. Petersburg, Russia. He also painted three copies for Johann Benedict Listing, Wilhelm Eduard Weber, and Wolfgang Sartorius von Waltershausen. Listing died 1882 in Göttingen. His copy was bought in 1883 by the National Gallery Berlin from Listing's widow and was donated in 1888 to the Prussian Academy of Science by the "Ministerium der geistlichen, Unterrichts- und Medicinalangelegenheiten." It is now owned by the Berlin-Brandenburg Academy of Sciences and Humanities.

200. Source: https://en.wikipedia.org/wiki/Carl_Friedrich _Gauss#/media/File:Carl_Friedrich_Gau%C3%9F_signature.s vg. "Carl Friedrich Gauss signature" by derivative work: Pbroks13 (talk) Carl Friedrich Gauss (1777-1855) - Carl_Friedrich_Gauss_Namenszug_von_1794.jpg. Licensed under Public Domain via Commons.

201. See Wikipedia "Constructible Polygon."

202. [Dunnington, 2004]. See Wikipedia "Carl Friedrich Gauss," including the German version.

203. For a detailed discussion of the history of area computation and the related operations, see Chapter 1 of [Dunham, 1990]. Here, we just summarize how the various operations can be carried out. Addition and subtraction are directly done with the compass. Multiplication, division, and taking of square root rely on the following drawing.
It has a baseline consisting of a and b, where in turn a is composed

of r and d. The semicircle has radius r. The line segment c forms a right angle with the baseline. The dashed line segment e connects two points of the halfcircle as shown. The second dashed line goes from the midpoint of e to the center of the semicircle, and thus forms a right angle with e.

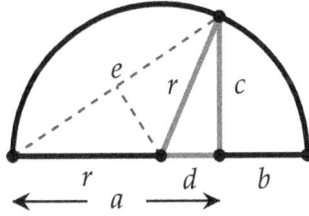

(Source: "Squared rectangle" by K. Truemper, released into Public Domain under Creative Commons CC0.)

By Pythagoras's theorem for the solid triangle with side lengths c, d, and r, we have $r^2 = c^2 + d^2$. Using $r = \frac{a+b}{2}$ and $d = a - r = \frac{a-b}{2}$, we have $r^2 = (a^2 + 2ab + b^2)/4 = c^2 + (a^2 - 2ab + b^2)/4$. Reduction of the last equation yields $c^2 = ab$.

For any two values of a, b, and c, the third value can be obtained via the above construction, as follows:

Given a and b: Without loss of generality, assume $a \geq b$. Determine $r = \frac{a+b}{2}$, draw the semicircle, draw c at right angle to base line.

Given a and c: Draw a with c at right angle, then add dashed segment e. Find the midpoint of e, erect the second dashed line at right angle, get the centerpoint of the semicircle, draw the semicircle, and thus get b. Note: The drawing is applicable only if $a \geq c$. If $a < c$, the above construction rules still apply, but a different drawing results where d is part of b instead of a.

Case of given b and c follows by symmetry with a and c.

Taking square root $\sqrt{x} = z$: Define $a = x$ and $b = 1$, then determine c. Since $x \cdot 1 = c^2$, we have $z = c$.

Multiplication $xy = z$: Use $a = x$ and $b = y$ to get c for which $c^2 = ab$. Use this c and $a = 1$ to get b for which $c^2 = 1 \cdot b$, which is the desired solution z.

Division $\frac{x}{y} = z$: Rewrite as $x = yz$, with x and y known. Determine $\sqrt{x}$. Define $a = y$ and $c = \sqrt{x}$, then determine b, which is the desired z.

204. Source: "Complex roots providing 17-sided heptadecagon" by K. Truemper, released into Public Domain under Creative Commons CC0.

205. Source: "Heptadecagon" by K. Truemper, released into Public Domain under Creative Commons CC0.

206. The only cases of interest have p odd. Otherwise, $p = 2^k \cdot q$ for some $k \geq 1$ and q odd, and a construction for the case of q immediately yields a construction for p as described earlier.

207. See Wikipedia "Complex plane" for the multiplication rules of the complex plane, which imply that the roots of $x^p - 1$ must be evenly distributed on the circle of the complex plane centered at the origin and with radius 1.

208. Gauss writes just $\frac{x^p-1}{x-1}$ when he actually means $\frac{x^p-1}{x-1} = 0$.

209. Gauss relies on the fact that $\frac{x^p-1}{x-1}$ is a polynomial whose roots are the $p - 1$ complex roots of the polynomial $x^p - 1$. This fact is a consequence of the *fundamental theorem of algebra*. See Wikipedia "Fundamental theorem of algebra" about details and the complicated history of its proofs, including four proofs by Gauss.

210. Letter to Gerling in 1819; see [Archibald, 1920]. Gauss had focused on the first complex root encountered as one traverses the unit circle counterclockwise from the single real root, which has value 1. That complex root is $R = \cos(\frac{2\pi}{17}) + \sin(\frac{2\pi}{17})i$ of the diagram. He found a way to compute that root using just the four basic arithmetic operations and taking of square root. Indeed, he established the following equation for the cosine of $\frac{2\pi}{17}$, written here in modern notation: $\cos\frac{2\pi}{17} = \frac{1}{16}[-1 + \sqrt{17} + \sqrt{34 - 2\sqrt{17}} + 2\sqrt{17 + 3\sqrt{17} - \sqrt{34 - 2\sqrt{17}} - 2\sqrt{34 + 2\sqrt{17}}}]$. Evaluation of the formula just requires repeated application of the five operations of addition, subtraction, multiplication, division, and the taking of square root. For any angle α, $\sin\alpha = \sqrt{1 - \cos^2\alpha}$, so $\sin\frac{2\pi}{17}$ can then be computed by these five operations as well. Thus, the point $R = \cos(\frac{2\pi}{17}) + \sin(\frac{2\pi}{17})i$ of the complex plane can be constructed.

211. See Wikipedia "Fermat number" for the interesting history of these primes. Chapter 10 of [Dunham, 1990] has details of Euler's proof that $2^{(2^5)} + 1 = 4,294,967,297$ is equal to $641 \cdot 6,700,417$ and thus is not a prime, at the time an astonishing achievement.

212. Source: https://en.wikipedia.org/wiki/Pierre_de_Ferm at#/media/File:Pierre_de_Fermat.jpg. "Pierre de Fermat" by http://www-groups.dcs.st-and.ac.uk/~history/PictDisplay/F

ermat.html. Licensed under Public Domain via Commons.

213. See Wikipedia "Pierre Wantzel."

214. For example, the *heptagon*, with 7 edges, and the *nonagon*, with nine edges, are regular polygons that cannot be constructed. Indeed, 7 is not a Fermat prime, and 9 is equal to $3 \cdot 3$ and thus the product of two identical Fermat primes.

215. See Wikipedia "Carl Friedrich Gauss."

216. Source: https://de.wikipedia.org/wiki/Carl_Friedrich_G au%C3%9F#/media/File:Braunschweig_Gauss-Denkmal_17-ecki ger_Stern.jpg. "Braunschweig Gauss-Denkmal 17-eckiger Stern" by Benutzer:Brunswyk. Licensed under CC BY-SA 3.0 via Creative Commons.

217. Source: https://en.wikipedia.org/wiki/Paolo_Ruffini# /media/File:Ruffini_paolo.jpg. "Ruffini Paolo" by Unknown. Licensed under Public Domain via Commons.

218. Source: https://en.wikipedia.org/wiki/Niels_Henrik_A bel#/media/File:Niels_Henrik_Abel.jpg. "Niels Henrik Abel" by Johan Gørbitz - Originally uploaded to English wikipedia by en:User:
Pladask. Licensed under Public Domain via Commons.

219. See Wikipedia "Cubic function."

220. See Wikipedia "Quartic Function."

221. See Wikipedia "Paolo Ruffini," "Niels Henrik Abel," "Évariste Galois," and "Pierre Wantzel."

222. Source: https://commons.wikimedia.org/wiki/File:Evar iste_galois.jpg. Portrait by Unknown was owned by Nathalie-Théodore Chantelot, E. Galois's older sister, and her daughter Mrs. Guinard. It was released by Paul Dupuy, École Normale Supérieure professor of history, with his article "La vie d'Évariste Galois," in 1896. "Evariste Galois" by Unknown - Iyanaga, Shokichi, Springer-Verlag Tokyo, 1999. http://www.win.tue.nl/~aeb/at/GaloisCorr espondence.html. Licensed under Public Domain via Commons.

223. A detailed account is given in [Livio, 2005].

224. See Wikipedia "Group (mathematics)."

NOTES

225. The binary field is an example of a finite field. Indeed, it is the smallest field, defined as follows.

1. The numbers are 0 and 1.

2. Addition: $0 + 0 = 0, 0 + 1 = 1 + 0 = 1, 1 + 1 = 0$.

3. Subtraction: $0 - 0 = 0, 0 - 1 = 1 - 0 = 1, 1 - 1 = 0$.

4. Multiplication: $0 \cdot 1 = 1 \cdot 0 = 0, 1 \cdot 1 = 1$.

5. Division: $0/1 = 0, 1/1 = 1$.

For details about finite fields, see Wikipedia "'Finite field."

226. See Wikipedia "Finite field."

227. See Wikipedia "Pierre Wantzel" and "Angle trisection." A summary of Wantzel's arguments is included in [Cajori, 1918].

228. Assume that trisection can be done for any angle. Use trisection to divide the 360 degrees of the circle into three 120 degree slices. This was already done by the ancient Greeks via construction of the equilateral triangle.

(Source: "Trisection of circle" by K. Truemper, released into Public Domain under Creative Commons CC0.)
To each 120 degree slice, apply trisection again, getting a total of nine 40 degree slices. From these, we directly derive the nonagon with nine edges. But we have seen that this polygon cannot be constructed, so it must be impossible to carry out the second trisection step.

229. See Wikipedia "Galois theory."

230. See Wikipedia "Doubling the cube" and "Pierre Wantzel." For a summary of the arguments, see [Cajori, 1918].

231. See Wikipedia "Galois theory."

232. Source: "Triangle area" by K. Truemper, released into Public Domain under Creative Commons CC0.

233. The subdivision into triangles is carried out as follows.

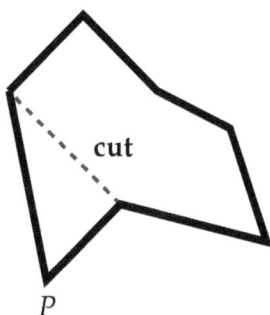

(Source: "Triangulation of polygon" by K. Truemper, released into Public Domain under Creative Commons CC0.)

Define an *ear* of the polygon to be a vertex P such that the line segment connecting the two neighbors of P lies entirely in the interior of the polygon. The two ears theorem states that a polygon has at least two ears; see Wikipedia "Two ears theorem" for details and proof. Evidently the line segment and the two edges attached to an ear P form a triangle that contains no other part of the polygon. Hence the triangle can be cut off, resulting in a polygon with one less vertex. Repeat until the polygon has become a triangle.

234. Details are provided on pp. 17–19 [Dunham, 1990]. Here is a short version.

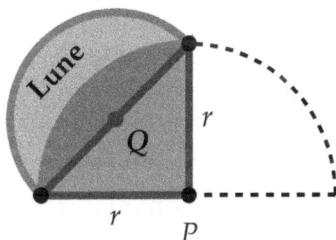

(Source: "Lune and triangle" by K. Truemper, released into Public Domain under Creative Commons CC0.)

We begin with the quarter circle that is centered at P and has radius r. It consists of a region we call the lens and a triangle where the two shorter sides form a right angle and are radii of the quarter circle. The third and longer side of the triangle is the diameter of the semicircle centered at Q and consisting of the lune and the lens. By Pythagoras's theorem, the longer side of the triangle has length $\sqrt{r^2 + r^2} = r\sqrt{2}$. This implies that the areas of the quarter circle and the semicircle are the same; the value is $\frac{1}{4}r^2\pi$. Removing the lens they have in common, the area of the lune must be equal to that of the triangle.

235. Source: "Polygon area" by K. Truemper, released into Public Domain under Creative Commons CC0.

236. Source: "Lune of Hippocrates" by K. Truemper, released into Public Domain under Creative Commons CC0.

237. "Archimedes parabola with triangle" by K. Truemper, released into Public Domain under Creative Commons CC0.

238. See Wikipedia "Archimedes."

239. See Chapter 4.

240. [Cajori, 1918] has a summary.

241. See Chapter 2.

242. Euclid's postulates are as follows:

   1. A straight line segment can be drawn connecting any two points.

   2. Any straight line segment can be extended indefinitely in a straight line.

   3. A circle can be drawn around any point and with any radius.

   4. All right angles are equal to one another.

   5. (The parallel postulate) If a straight line intersects two straight lines such that the sum of the inner angles on one side is less than two right angles, then the two lines must intersect on that side when extended far enough.

See Wikipedia "Euclidean geometry" for further details.

243. See Wikipedia "Playfair's axiom." Note that the first four axioms imply that there is at least one parallel line. Thus, the condition of Playfair's axiom that there is *at most* one parallel line means that there is *exactly* one such line.

244. See Wikipedia "Parallel postulate."

245. Source: https://commons.wikimedia.org/wiki/File:John _Playfair_by_Sir_Henry_Raeburn.jpg. "John Playfair" by Henry Raeburn. Public Domain under US copyright code PD-old-100.

246. An axiom C is implied by other axioms if C holds whenever the other axioms hold.

247. See Wikipedia "Parallel postulate."

248. [Engel and Stäckel, 1895] and [Königliche Gesellschaft der Wissenschaften, Göttingen, 1900].

249. Source: https://de.wikipedia.org/wiki/Ferdinand_Karl _Schweikart#/media/File:Ferdinand_Karl_Schweikart.jpg. "Ferdi-nand Karl Schweikart" by Unknown - http://www.liveinternet.r u/users/kakula/post153809122/. Licensed under Public Domain under US copyright code PD-old-70.

250. p. 244 [Engel and Stäckel, 1895].

251. p. 246 [Engel and Stäckel, 1895].

252. p. 249, 250 [Engel and Stäckel, 1895] contains the letter of Gauss to Taurinus: "About your attempt I have little to comment except that it is incomplete. ... I guess that you have not worked for a long time on this topic. For me, it has been over 30 years, and I do not believe, that anybody has been more involved than me with the second part [concerning the sum of angles of any triangle less than 180 degrees], although I have not published anything about this. The assumption that the sum of the 3 angles is smaller than 180 degrees results in a particular, from present-day (Euclidean) different, geometry which I have worked out to my satisfaction, so that I can answer every question, except for determination of a constant which cannot be determined a priori." Gauss's letter con-cludes: "At any rate, you must view the above material as private information that in no way should be published or used in such a way that it could become public. Perhaps when I have more time than available at present, I will make these results public."

253. Source: https://en.wikipedia.org/wiki/Nikolai_Lobach evsky#/media/File:Lobachevsky.jpg. "Lobachevsky" by Lev Dmitrie-vich Kryukov. Licensed under Public Domain via Commons.

254. Source: https://en.wikipedia.org/wiki/Henri_Poincar%C3 %A9#/media/File:Henri_Poincar%C3%A9-2.jpg. "Henri Poincaré-2" by Unknown - http://www.mlahanas.de/Physics/Bios/image s/HenriPoincare.jpg. Licensed under Public Domain via Com-mons.

255. Source: "Poincaré disk" by K. Truemper, released into Public Domain under Creative Commons CC0.

256. Indeed, it is easy to see that there are an infinite number

of lines going through point P that are parallel to line A. Thus, Lobachevsky's condition that there are at least two parallel lines implies that, for any given line and any point not on that line, there are an infinite number of parallel lines going through that point.

257. Source: http://www.titoktan.hu/Bolyai_a.htm. "János Bolyai" photo by copyright holder Támas Dénes, who has kindly granted permission to use the photo. According to [Dénes, 2011] and the Wikipedia "János Bolyai" entry, no original portrait of Bolyai survives, and an unauthentic picture appears in some encyclopedias and on a Hungarian postage stamp.
The relief of János Bolyai shown here is part of six reliefs in front of the Culture Palace in Marosvásárhely, Romania. According to the investigation reported in [Dénes, 2011] that included computer simulation using images across three generations, the relief likely is a good, indeed the only authentic, representation of the mathematician.

258. See Wikipedia "Non-Euclidean geometry."

259. See Wikipedia "János Bolyai."

260. Gauss's evaluation is confirmed by recent investigations into the work of János Bolyai; see [Dénes, 2011] and [Kiss, 1999].

261. Gauss's letter to Taurinus includes the following philosophical statement, see p. 250 [Engel and Stäckel, 1895]:
"... All efforts of mine to find a contradiction, an inconsistency in this non-Euclidean geometry have been fruitless, and the only thing that makes our brain resist it [the geometry], is that, if true, there would have to be a certain (though not known by us) constant of linearity. But I think that, despite the vacuous word-wisdom of the metaphysicists, we know not enough or indeed nothing about the true nature of space, that we can permit us to confuse something that appears to us unnatural with something that is absolutely impossible. If the Euclidean geometry was the true one, and the constant [used in the non-Euclidean geometry] was related to values that could be measured on earth or up in the sky, then we could determine the true nature [of space] *a posteriori*. So jokingly I have sometimes expressed the wish that the Euclidean geometry was not the true one, since then we would have an absolute measure [of the constant] a priori."

262. Source: http://www-groups.dcs.st-and.ac.uk/history/ PictDisplay/Beltrami.html. "Eugenio Beltrami" by Unknown.

Public Domain under US copyright code PD-old-70.

263. See Wikipedia "Eugenio Beltrami."

264. See Wikipedia "Elliptic geometry."

265. Source: `https://en.wikipedia.org/wiki/Elliptic_geometr y#/media/File:Triangles_%28spherical_geometry%29.jpg`. "Triangles (spherical geometry)" by Lars H. Rohwedder, Sarregouset - Own work from source files Image:OgaPeninsulaAkiJpLandsat.jpg (GFDL) and Image:Orthographic Projection Japan.jpg (GFDL and CC-By-SA). Licensed under CC BY-SA 3.0 via Commons.

266. See Wikipedia "Great circle."

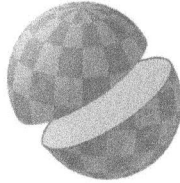

(Source: `https://en.wikipedia.org/wiki/Great_circle#/medi a/File:Great_circle_hemispheres.png`. "Great Circle" by Jhbdel, en.wikipedia. Licensed under CC BY-SA 3.0 via Commons.)

267. Technically, any two antipodal points are *identified* to represent one point of the geometry. This is needed so that Euclid's axiom is satisfied that two points uniquely determine a line. See Wikipedia "Elliptic geometry."

Chapter 6 Proof

268. For example, the website snopes.com for "urban legends, folklore, myths, rumors, and misinformation" can be helpful.

269. See Wikipedia "Occam's razor."

270. Source: `https://en.wikipedia.org/wiki/Albert_Einstein# /media/File:Albert_Einstein_%28Nobel%29.png`. "Albert Einstein (Nobel)" by Unknown - Official 1921 Nobel Prize in Physics photograph. Licensed under Public Domain via Commons.

271. Source: `https://en.wikipedia.org/wiki/Minkowski_space# /media/File:GPB_circling_earth.jpg`. "GPB circling earth" by NASA - `http://www.nasa.gov/mission_pages/gpb/gpb_012.html`. Licensed under Public Domain via Commons.

272. See Wikipedia "Space-time."

273. Source: https://www.math.ubc.ca/~cass/Euclid/ybc/ybc.html. Permission for use granted by William A. Casselman, photographer and copyright holder. The clay tablet is part of the Yale Babylonian Collection. Provenance unknown, dated approximately 1800–1600 BCE. Purchased around 1912 by an agent of J. P. Morgan, who contributed it to Yale University as part of the foundation of its Babylonian Collection.

274. Source: https://en.wikipedia.org/wiki/File:Plimpton_3 22.jpg. "Plimpton 322" by photo author unknown. Licensed under Public Domain via Commons.

275. See Wikipedia "Pythagorean theorem" for various proofs.

276. p. 215 [Rudman, 2007].

277. [Ossendrijver, 2016].

278. p. 21 [Wilson, 1828].

279. See Wikipedia "Apparent retrograde motion."

280. Source: "Jupiter velocity graph" by K. Truemper, released into Public Domain under Creative Commons CC0. The drawing is based on a figure of [Ossendrijver, 2016].

281. Source: Photo by Mathieu Ossendrijver, Humboldt University/British Museum. M. Ossendrijver has kindly granted permission for use of the photo.

282. [Ossendrijver, 2016].

283. See Wikipedia "Aristotle."

284. See Wikipedia "Euclid."

285. Source: https://commons.wikimedia.org/wiki/File:Aristo tle_Altemps_Inv8575.jpg. "Aristotle Altemps Inv8575" by Copy of Lysippus - Jastrow (2006). Licensed under Public Domain via Commons.

286. Here are some of the accomplishments of Archimedes; for details see [Netz and Noel, 2007]:

1. An iterative scheme to compute π with as high a precision as desired, using approximating polygons; he carried out that

method for the 96-sided polygon, getting the bounds $3\frac{10}{71} <$ $\pi < 3\frac{1}{7}$, with midpoint estimate 3.14185. The estimate matches three digits after the decimal point of the actual value $\pi =$ 3.14159 . . . .

2. Computation of areas and volumes with a precursor of integral calculus, see Chapters 3 and 5.

3. Schematics for geometric problem that do not correspond to reality but show logic relationships of points and lines.

4. Counting of geometric configurations.

287. For the period up to about 100 BCE, any list should include Thales of Miletus, Pythagoras of Samos, Zeno of Elea, Hippocrates of Chios, Archytas of Tarentum, Theaetetus of Athens, Eudoxus of Cnidus, Apollonius of Perga, and Hipparchus of Nicaea. Wikipedia has detailed information for each mathematician of the list.

288. [Lenzen, 2004].

289. Leibniz anticipated results for the following areas of modern logic.

1. *Propositional logic*: It uses *propositional variables* with values *True* and *False* to encode elementary facts and expresses their relationships with the basic *operators* "not," "and," "or," plus advanced operators such as "if . . . then" and "only if." For example, let the variable x have value *True* if it is raining, and define variable y to have that value if a person uses an umbrella. Then the formula *if x then y* has value *True* if it is not raining or the person uses an umbrella.

2. *First-order logic*: It uses *predicates*, also called *truth functions*, in addition to propositional variables. An example is the function *woman*(x) where x represents an arbitrary person of the human population and *woman*(x) has value *True* if x is a woman and *False* otherwise. For the description of relationships among predicates, the operators of propositional logic are used plus the *quantifiers* "exists" and "for all." For example, the compact statement *exists x woman*(x) says that there is a person on earth who is a woman. For details, see Wikipedia "First-order logic."

3. *Modal logic*: It generalizes first-order logic by allowing operators such as "possibly," "necessarily," "impossibly," "it is obligatory that," and "it is permissible that." For example, the statement "The woman possibly was an airline employee"

can be expressed in modal logic. For details, see Wikipedia "Modal logic."

290. [Peckhaus, 1997].

291. See Wikipedia "Calculus ratiocinator."

292. See Wikipedia "Characteristica universalis."

293. See Wikipedia "Artificial intelligence."

294. See Wikipedia "Calculus ratiocinator" and "Characteristica universalis."

295. See [Morgan, 1847] and p. 1863, 1864 [Newman, 1956].

296. Source: `https://en.wikipedia.org/wiki/File:De_Morgan_Augustus.jpg`. "Memoir of Augustus de Morgan" by Sophia Elizabeth De Morgan, 1882. Licensed under Public Domain via Commons.

297. See pp. 1864–1868 [Newman, 1956]. Boole's approach may be sketched as follows.
He starts with classes having certain features. Let x, y, z denote such classes. The class containing all features, the *universe of discourse*, is denoted by 1, and that class not containing any feature, the *null class*, is denoted by 0. The class having the features of both x and y is xy. For x and y with disjoint features, $x + y$ is the class where each element has either the features of x or those of y. Finally, if $z = x + y$, then $x = z - y$ is declared to hold.
Since 1 is the universe of discourse and 0 is the null class, $1x = x$ and $0x = 0$. The complement of x is $1 - x$. These definitions cause problems: For example, $x + x$ makes sense only if x is the empty class, and $x - y$ is defined only if x contains y. But then Boole defines a specialized version by adding the condition that each class must be equal to 0 or 1. Boole handles the undefined $x + x$ by allowing integers larger than 1 during computations. The latter change has the drawback that intermediate steps of calculations may not have an interpretation in terms of classes. Nevertheless, Boole's system makes reliable logic calculations possible.

298. Source: `https://en.wikipedia.org/wiki/George_Boole#/media/File:George_Boole_color.jpg`. "George Boole color" by Unknown. Licensed under Public Domain via Commons.

299. Source: `https://archive.org/`. Search for "Boole Laws of

Thought." Licensed under Public Domain Mark 1.0 via Creative Commons.

300. p. 69, 70 [Boole, 1854].

301. The exclusive "or" was replaced by the inclusive version, and the subtraction operator was dropped.

302. See Wikipedia "Boolean algebra."

303. See Wikipedia "Boolean algebra."

304. [Frege, 1879].

305. Besides propositional logic and first-order logic cited earlier, Frege also defined *second-order logic* where quantifiers "exists" and "for all" may not only refer to a universe of discourse, but may apply to predicate functions. One may re-express this by saying that second-order logic allows quantification over sets instead of just individual members of the universe. An example statement is *for every subset P of the universe and every member x of the universe, x is in P or x is not in P.*

306. Source: https://en.wikipedia.org/wiki/Gottlob_Frege#/media/File:Young_frege.jpg. "Young Frege" by Unknown. Licensed under Public Domain via Commons.

307. Source: https://en.wikipedia.org/wiki/Begriffsschrift#/media/File:Begriffsschrift_Titel.png. Uses digitized version at http://gallica.bnf.fr/ark:/12148/bpt6k65658c. Licensed under Public Domain via Commons.

308. [Frege, 1893].

309. [Frege, 1884].

310. p. 1 [Frege, 1893].

311. Source: https://archive.org/. Search for "Gottlob Frege Grundgesetze der Arithmetik." Public Domain.

312. The appendix opens with the statement, "To a science writer there is hardly anything as undesirable as the realization at the end of a project that one of the foundations of the building has been shattered." It concludes with, "As fundamental problem of arithmetic one can view to be the question: How do we capture the logic elements, in particular the numbers? Why are we justified to interpret the numbers as things? Even though that problem is

not solved to the extent as I had thought during the writing of this volume, I still have no doubt that the road to success has been identified."

313. Source: `https://en.wikipedia.org/wiki/Bertrand_Russe ll#/media/File:Russell_in_1938.jpg`. "Bertrand Russell" by Unknown. Licensed under Public Domain via Commons.

314. The following discussion of the contradiction inherent in Frege's set definition is based on Wikipedia "Russell's paradox." Define a set to be *normal* if it does not contain itself as an element, and to be *abnormal* otherwise. For example, the set of spoons is itself not a spoon, and thus is normal. On the other hand, the set where each element is not a spoon, itself isn't a spoon either, and thus is abnormal.

Now define S to be the set of all normal sets. Is S normal or abnormal? Suppose S is normal and thus does not contain itself as an element. But that contradicts that S contains all normal sets. Now suppose S is abnormal and thus contains itself as an element. But by definition of S, all elements of S are normal. Thus S is normal, another contradiction.

Since both cases result in a contradiction, the set S cannot exist. But Frege's construction of sets via logic conditions declares S to exist, which can only mean that the construction is inherently faulty.

315. See Wikipedia "David Hilbert."

316. See Wikipedia "Hilbert's problems" for details, including current status of solved versus still unsolved.

317. Source: `https://en.wikipedia.org/wiki/David_Hilbert#/m edia/File:Hilbert.jpg`. "David Hilbert" by Unknown. Licensed under Public Domain via Commons.

318. The definition of well-ordering is based on a feature of the set N of natural numbers: Take any subset S of N containing at least one number; then among the numbers of S, one number is smallest. Based on that feature of N, a general set with a given ordering is said to be *well ordered* if any nonempty subset has a smallest element.

Now some sets are not well ordered, for example, the set of integers with their natural ordering. Indeed, the subset consisting of -1, -2, $-3\ldots$ has no smallest element. But the integers *can* become well ordered when we use a different ordering. For example, we can declare the integers to have the increasing order $0, -1, 1, -2, 2,$

−3, 3 . . ., where we are suspending the usual interpretation of the numbers. Then every subset does have a smallest element.

319. Source: `https://en.wikipedia.org/wiki/Ernst_Zermelo#` `/media/File:Ernst_Zermelo.jpeg`. "Ernst Zermelo" by Konrad Jacobs - `http://owpdb.mfo.de/detail?photo_id=8666`. Licensed under CC BY-SA 2.0 de via Commons.

320. Source: `https://en.wikipedia.org/wiki/Basket#/media/` `File:Gullah_basket.JPG`. "Gullah basket" by Bubba73 (Jud Mc-Cranie) - Own work. Licensed under CC BY-SA 4.0 via Commons.

321. See Wikipedia "Axiom of countable choice."

322. Source: `https://en.wikipedia.org/wiki/Abraham_Fraenk` `el#/media/File:Adolf_Abraham_Halevi_Fraenkel.jpg`. "Adolf Abraham Halevi Fraenkel" by The David B. Keidan Collection of Digital Images from the Central Zionist Archives (via Harvard University Library). Licensed under Public Domain via Commons.

323. See Wikipedia "Banach-Tarski paradox."

324. Source: `https://en.wikipedia.org/wiki/Banach%E2%80%9` `3Tarski_paradox#/media/File:Banach-Tarski_Paradox.svg`. "Banach-Tarski Paradox" by Benjamin D. Esham. Licensed under Public Domain via Commons.

325. Source: Institute of Mathematics of the Polish Academy of Sciences, which kindly granted use of the photo.

326. Source: `http://owpdb.mfo.de/detail?photo_id=6091`. "AlfredTarski1968" by George M. Bergman - The Oberwolfach photo collection. Licensed under Commons CC BY-SA 2.0 DE.

327. See Wikipedia "J. E. L. Brouwer."

328. Source: "BrKop7_3.jpg" photo of Brouwer Archive, National Archive Haarlem. The Archive kindly has granted use of the photo.

329. See Wikipedia "Topology."

330. The theorem says that, for any compact convex set S in any n-dimensional Euclidean space and any continuous function f that maps S into itself, there is a point $x \in S$ such that $f(x) = x$. The point x is called a *fixed point* of f. For details and other fixed-point theorems, see Wikipedia "Brouwer's fixed-point theorem" and "Fixed-point theorem."

331. See Wikipedia "Fixed-point theorem."

332. See Wikipedia "Intuitionism."

333. Aristotle's law of the excluded middle states that, for any proposition, either the proposition is true or it is false. Thus, there is no third case possible.

334. Here are two additional examples based on discussion on p. 47, 48 [Weyl, 1921], reprinted p. 165, 166 [Thiel, 1982]:
Suppose we have established an infinite collection of real numbers x, all of which are less than some given number, say 1. The numbers x may not be directly given, but indirectly specified by some process or theorem. The axiomatic method then allows the claim that there is a real number, say y, with the following property: All numbers x are less than or equal to y, and there is no number z smaller than y for which this can be claimed. The number y is called the *least upper bound* for the x numbers.
The intuitionist will not accept the claim or the use of such y unless a constructive procedure is offered. In particular, the Dedekind cut of Chapter 2, which invokes such bounds y en masse while creating the real numbers, is rejected outright by the intuitionist.
As second example, the intuitionist rejects the claim that every infinite set contains a countable infinite subset. Since that result is a consequence of the axiom of countable choice—see Wikipedia "Axiom of countable choice"— that axiom as well as the general axiom of choice are rejected as well.

335. [Weyl, 1921], reprinted pp. 157–178 [Thiel, 1982]. See also Wikipedia "Hermann Weyl."

336. Source: https://en.wikipedia.org/wiki/Hermann_Weyl#/media/File:Hermann_Weyl_ETH-Bib_Portr_00890.jpg. "Hermann Weyl" by ETH Zürich - ETH-Bibliothek Zürich, Bildarchiv. Licensed under CC BY-SA 3.0 via Commons.

337. For details of the struggle, see Stanford Encyclopedia of Philosophy "Luitzen Egbertus Jan Brouwer."

338. When *ZF* is assumed, the axiom of choice can be used to prove the well-ordering theorem, and vice versa. As a result, *ZFC* is sometimes stated with the well-ordering theorem instead of the axiom of choice. For example, see Wikipedia "Zermelo-Fraenkel set theory."

339. See Wikipedia "L. E. J. Brouwer."

340. See Wikipedia "Hermann Weyl."

341. The entire section is based on Wikipedia "Principia Mathematica."

342. Source: HUP Whitehead, Alfred North (3b), W395042_1, Harvard University Archives. "Alfred North Whitehead" photo is part of collection of photos of Harvard faculty and buildings taken in 1936 by Richard Carver Wood. Harvard University kindly granted permission to use the photo.

343. Source: https://archive.org/. Search for "Principia Mathematica Whitehead Russell." Public Domain.

344. For example, the very first statement defines the implication *p implies q* by the statement *not p or q*.

345. See Wikipedia "Principia Mathematica."

346. In terms of logic: A systems of axioms is consistent if there is no statement S such both S and the negation of S can be proved.

347. Generally, completeness and consistency are difficult to establish. For example, Gauss hesitated to publish his results on hyperbolic geometry partly because he could not establish consistency; see Chapter 5. Beltrami then proved that hyperbolic geometry is consistent if and only if this is true for Euclidean geometry. This relative result was replaced by absolute statements about consistency of Euclid's geometry, and thus of hyperbolic geometry, in the 20th century; see Wikipedia "Euclidean geometry."

348. Source: https://en.wikipedia.org/wiki/Emil_du_Bois-Rey mond#/media/File:Emil_DuBois-Reymond_BNF_Gallica_crop.jpg. "Emil du Bois-Reymond" by Haase phot. Berlin. Upload, stitch and restoration by Jebulon - Bibliothèque Nationale de France. Public Domain.

349. See Wikipedia "Ignoramus et ignorabimus."

350. See Wikipedia "Ignoramus et ignorabimus." Hilbert read on the German radio an abbreviated four-minute version of the speech. The recording is available at http://www.maa.org/press/periodic als/convergence/david-hilberts-radio-address, together with the German text as well as an English translation. The speech concludes with the cited passage.

351. Source: https://en.wikipedia.org/wiki/David_Hilbert#/m

edia/File:G%C3%B6ttingen_Stadtfriedhof_Grab_David_Hilbert. jpg. "Hilbert Grab" by Kassandro - Own work, https://commons. wikimedia.org/w/index.php?curid=4219496. Licensed under CC BY-SA 3.0 via Commons.

352. See Wikipedia "Hilbert's program" for details.

353. We cannot include even a summarizing discussion of the rules of finitary proofs as proposed by Hilbert and later amended by others. The principal idea is that the proof steps cannot invoke infinite processes. But formalizing this idea is another matter, and to date no complete set of rules for finitary proofs has been stated. For details about the various viewpoints, see Stanford Encyclopedia of Philosophy "Hilbert's program."

354. See Stanford Encyclopedia of Philosophy "Hilbert's program."

355. Source: https://en.wikipedia.org/wiki/Kurt_G%C3%B6del# /media/File:Kurt_g%C3%B6del.jpg. "Kurt Goedel, ca. 1926" by Unknown. Licensed under Public Domain via Commons.

356. For references and details, see Stanford Encyclopedia of Philosophy "Gödel's Incompleteness Theorems."

357. This condition is not exactly the same as in the first incompleteness theorem. Indeed, it is a bit stronger than in the first incompleteness theorem, where the condition is very weak.

358. It may appear that the condition "contains a certain amount of elementary arithmetic" may be severe enough that a number of mathematical systems are not affected by the two incompleteness theorems. But that condition can be weakened as long as the system contains parts that can be *interpreted* as a certain amount of elementary arithmetic.

359. See Chapter 4.

360. Source: https://de.wikipedia.org/wiki/Datei:Max_Plan ck_%281858-1947%29.jpg. "Max Planck" by Unknown - http: //www.sil.si.edu/digitalcollections/hst/scientific-ide ntity/CF/display_results.cfm?alpha_sort=p. Licensed under Public Domain via Commons.

361. One can express the two conditions for independence in the terminology of theorems as follows. Requiring consistency of system S with axiom A added is equivalent to demanding that $A =$ false cannot be proved to be a theorem of S. Calling for consistency

of S with the negation of A added is the same as demanding that $A =$ true cannot be proved to be a theorem of S.

362. See Stanford Encyclopedia of Philosophy "Kurt Gödel."

363. See Wikipedia "Paul Cohen."

364. Source: Courtesy Cohen Family/Stanford University, who kindly granted permission to use the photo.

365. See Wikipedia "Paul Cohen." It may seem odd that Gödel compared the effect the proof had on him with his reaction to a really good play. But brain science has shown that the area of the brain evaluating the beauty of mathematical statements is also active when responding to visual, musical, and even moral beauty; see [Zeki et al., 2014] and Chapter 13.

366. See Wikipedia "Forcing (mathematics)."

367. [Wolchover, 2013].

368. See Wikipedia "Nicolas Bourbaki." Professor Bourbaki supposedly works at the University of Nancago, a name composed from *Nancy*, France, and *Chicago*, USA. One is tempted to guess that Nancago is situated halfway between these two cities and thus in the middle of the Atlantic Ocean. Maybe it is the Atlantis of lore?

369. [Bourbaki, 1948].

370. See Wikipedia "Fermat's last theorem."

371. See Wikipedia "Fermat's last theorem."

372. See Wikipedia "Fermat's last theorem" for details of the historical developments.

373. See Wikipedia "Andrew Wiles."

374. See Wikipedia "Modularity theorem."

375. Source: https://en.wikipedia.org/wiki/Andrew_Wiles#/media/File:Andrew_wiles1-3.jpg. "Andrew Wiles" copyright C. J. Mozzochi, Princeton N.J - http://www.mozzochi.org/deligne60/Deligne1/_DSC0024.jpg. The copyright statement grants free use.

376. See Wikipedia "Modularity theorem."

377. [Singh, 1997].

378. An effort is underway to start construction of the computer. See Wikipedia "Analytical engine."

Chapter 7 Computing Machines

379. See Wikipedia "Abacus."

380. Source: https://en.wikipedia.org/wiki/Abacus#/media/Fi le:Boulier1.JPG. "Abacus" by HB - Own work, Public Domain.

381. Source: https://en.wikipedia.org/wiki/Gear#/media/Fi le:Interactive_gears_in_the_hind_legs_of_Issus_coleop tratus_from_Cambridge_gears-3.jpg. "Issus coleoptratus" by University of Cambridge (Malcolm Burrows & Gregory Sutton) - http://www.cam.ac.uk/research/news/functioning-mechanic al-gears-seen-in-nature-for-the-first-time. Licensed under CC BY-SA 3.0 via Commons.

382. See Wikipedia "Gear."

383. See Wikipedia "Issus coleoptratus."

384. See Wikipedia "Antikythera mechanism," which contains details of the recovery and interpretation of the mechanism.

385. Source: https://en.wikipedia.org/wiki/Antikythera_mech anism#/media//File:Antikythera_model_front_panel_Mogi_Vic entini_2007.JPG. "Antikythera mechanism" by I, Mogi. Licensed under CC BY 2.5 via Commons.

386. A schematic of the gearing of the Antikythera mechanism is shown on the next page.
It includes the 2012 published interpretation of existing gearing, gearing added to complete known functions, and proposed gearing to accomplish additional functions, namely true sun pointer and pointers for the five then-known planets. For details and references, see Wikipedia "Antikythera mechanism."
(Source: https://en.wikipedia.org/wiki/Antikythera_mechanis m#/media/File:AntikytheraMechanismSchematic-Freeth12.png. "Antikythera Gears" by SkoreKeep - Own work. Licensed under CC BY-SA 3.0 via Commons.)

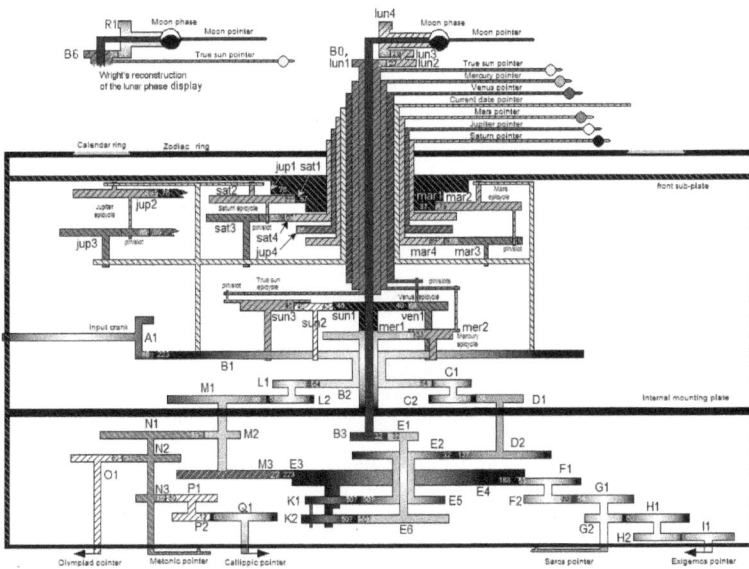

387. See Wikipedia "Blaise Pascal."

388. The 9-complement method proceeds as follows. Suppose y is to be subtracted from x. Derive from x the *9-complement* x' by replacing each digit d of x by $9 - d$. Add x' and y, say getting z. Then the 9-complement of z is the desired quantity $x - y$. For details and other complement methods converting subtraction to addition, see Wikipedia "Method of complements."

389. See Wikipedia "Pascal's calculator."

390. Source: https://en.wikipedia.org/wiki/Blaise_Pascal#/media/File:Arts_et_Metiers_Pascaline_dsc03869.jpg. "Pascaline" by David Monniaux, copyright 2005. Licensed under CC BY-SA 3.0 via Commons.

391. Source: https://en.wikipedia.org/wiki/Blaise_Pascal#/media/File:Blaise_Pascal_Versailles.JPG. "Blaise Pascal" by unknown. Copy of the painting by François II Quesnel, which was made for Gérard Edelinck in 1691. - Own work. Licensed under CC BY 3.0 via Commons.

392. Source: https://en.wikipedia.org/wiki/Pascal%27s_calculator#/media/File:Detail_of_the_pascaline%27s_carry_mecanism_-_the_sautoir.jpg. "Pascaline - detail" by Unknown - Oeuvres de Blaise Pascal, Chez Detune, La Haye, Public Domain.

In the input/output diagram for one digit, rotation of the horizontal input wheel in the top right-hand corner is transmitted via gears with pins as cogs to the horizontal shaft in the center. In turn, that shaft drives the output drum via another wheel with pins. The complicated mechanism in the center handles the carry function.

393. Source: https://en.wikipedia.org/wiki/Pascal%27s_cal culator#/media/File:Detail_of_the_pascaline%27s_carry_ mecanism_-_the_sautoir.jpg. "Pascaline - detail" by Unknown - Oeuvres de Blaise Pascal, Chez Detune, La Haye, Public Domain. The carry function is accomplished by separate wheels placed on the horizontal shafts. The drawing shows two such wheels coupled by a carry gadget that is hinged on the axis of the left-hand wheel. When the right-hand wheel is advanced from position 4 to 9, the carry gadget is lifted. When the right-hand wheel is then further advanced from 9 to 0, the carry gadget is released. It drops due to gravity and, in doing so, advances the left-hand wheel by 1. The ingenious use of gravity assures that multiple carry steps—for example, occurring when 1 is added to 9,999—are reliably done in a rapid cascade of individual steps, none of which require extra force by the operator.

394. See Wikipedia "Stepped reckoner."

395. See Wikipedia "Stepped reckoner."

396. Source: https://commons.wikimedia.org/wiki/File:Spro ssenrad_leibniz.png. "Sprossenrad Leibniz" von Gottfried Wilhelm Leibniz - Wilberg, Ernst Eberhard: 'Die Leibniz'sche Rechenmaschine und die Julius-Universität in Helmstedt', Braunschweig, 1977. Licensed under Public Domain via Commons.

397. Source: https://de.wikipedia.org/wiki/Sprossenrad#/m edia/File:Sprossenrad-rechner.jpg. "Modell einer Sprossen-radrecheneinheit" photo of Technische Sammlungen der Stadt Dresden. Licensed under CC BY-SA 3.0 via Commons.

398. See Wikipedia "Mechanical calculator."

399. See Wikipedia "Mechanical calculator."

400. Source: "Stepped reckoner" is photo 3,65:01 of the book *Das letzte Original, Die Leibniz-Rechenmaschine der Gottfried Wilhelm Leibniz Bibliothek* (The Last Original, the Leibniz Calculator of the Gottfried Wilhelm Leibniz Library) by Ariane Walsdorf, Klaus Badur, Erwin Stein, and Franz Otto Kopp, with preface by Georg Ruppelt.

Photograph: Maike Kandziora. Copyright 2014 Gottfried Wilhelm Leibniz Bibliothek - Niedersächsische Landesbibliothek, Hannover, Germany, which kindly granted permission to use the photo.

401. See Wikipedia "Leibniz wheel," which has an animation of the drawing.

402. Source: https://en.wikipedia.org/wiki/Leibniz_wheel#/media/File:Cylindre_de_Leibniz_anim%C3%A9.gif. "Cylindre de Leibniz animé" by Ezrdr - Own work. Licensed under CC BY-SA 3.0 via Creative Commons.

403. Source: "Input knobs, etc" is photo 3,65:59 of the book *Das letzte Original, Die Leibniz-Rechenmaschine der Gottfried Wilhelm Leibniz Bibliothek*. For details and license statement, see above citation.

404. Source: "Dial for multiplier" is photo 3,65:28 of the book *Das letzte Original, Die Leibniz-Rechenmaschine der Gottfried Wilhelm Leibniz Bibliothek*. For details and license statement, see above citation.

405. Source: "Carry mechanism" detail of photo 3,65:59 of the book *Das letzte Original, Die Leibniz-Rechenmaschine der Gottfried Wilhelm Leibniz Bibliothek*. For details and license statement, see above citation.

406. For extensive details about the operation of the calculator, see [Walsdorf et al., 2014].

407. See [Walsdorf et al., 2014] and [Badur and Rottstedt, 2004]. The latter reference contains a complete manual for operation of the machine.

408. For example, $12,405,897 \cdot 96,878,532 = 1,201,865,089,503,204$ is computed without any manual interference.

409. Leibniz's *Machina arithmeticae dyadicae* for binary addition and multiplication is described in De progressione dyadica, Pars II. That document also gives details of the machine for conversion of decimal to binary numbers. The Arithmeum, Bonn, has constructed a binary calculator based on Leibniz's description; see photo included in this chapter. Other constructions, including a binary version of the stepped reckoner, are described on pp. 232–245 [Walsdorf et al., 2014].

410. Source: "Leibniz Binary Calculator" photo by Anna Borutzky, copyright 2016 Arithmeum of the Rheinische Friedrich-Wilhelms-Universität, Bonn, Germany, which kindly granted permission to

use the photo.

411. See Wikipedia "Charles Babbage."

412. See Wikipedia "Dynamometer car."

413. Source: `https://en.wikipedia.org/wiki/Charles_Babbage#` `/media/File:Charles_Babbage_-_1860.jpg`. "Charles Babbage" by Unknown. Public Domain under US copyright code PD-old-70.

414. See Wikipedia "Flight recorder."

415. See Wikipedia "Difference engine," which contains a simple example where the difference engine computes values for a polynomial.

416. We have computed the estimate of modern-day cost as follows. According to p. 167 [Hyman, 1982], Babbage paid an outstanding draftsman, C. G. Jarvis, a very generous wage of one guinea per day. Jarvis had considerable technical knowledge and "deserves an honoured place in the history of the computer." One guinea is equivalent to £1.05; see Wikipedia "Guinea (British coin)." Assuming about 300 working days per year, yearly wages were therefore around £315. Today, in 2016, the job of draftsman no longer exists, but a first-rate engineer in an equivalent position may earn US$100,000 or more. Thus, we may use the factor $100,000/315 = 317$ to scale up Babbage's expenditure of £17,000 to an estimate of current cost of around five million US dollars.

417. See Wikipedia "Charles Babbage."

418. Source: `https://en.wikipedia.org/wiki/Charles_Babbag` `e#/media/File:Babbage_Difference_Engine.jpg`. "Difference Engine No. 2" photo by User:geni. Licensed under CC BY-SA 2.0 via Common.

419. See `http://www.computerhistory.org/babbage/`.

420. See Wikipedia "Analytical engine."

421. p. 166, 167 [Hyman, 1982].

422. An effort is underway to start construction of the computer. See Wikipedia "Analytical engine."

423. Source: `https://en.wikipedia.org/wiki/Analytical_Engin` `e#/media/File:AnalyticalMachine_Babbage_London.jpg`. "Trial model of Analytical Engine" by Bruno Barral (ByB). Licensed under

424. See Wikipedia "Ada Lovelace."

425. Source: https://en.wikipedia.org/wiki/Ada_Lovelace#/media/File:Ada_Lovelace.jpg. "Ada Lovelace" by Margaret Sarah Carpenter, 1836. Licensed under Public Domain via Commons.

426. See Wikipedia "Ada Lovelace." Below is the computer program Lovelace included in Note G. It is the first complex computer program ever published.

Diagram for the computation by the Engine of the Numbers of Bernoulli. See Note G. (page 722 et seq.)

(Source: https://en.wikipedia.org/wiki/Ada_Lovelace#/media/File:Diagram_for_the_computation_of_Bernoulli_numbers.jpg. "Diagram for computation by the Engine of the Numbers of Bernoulli" by Ada Lovelace - http://www.sophiararebooks.com/pictures/3544a.jpg. Licensed under Public Domain via Commons.)

427. See Wikipedia "Computer music." For examples of computer-composed music, go to http://www.computerhistory.org/atchm/algorithmic-music-david-cope-and-emi/.

428. Source: https://commons.wikimedia.org/wiki/File:Model_of_a_Turing_machine.jpg. "Turing machine" by GabrielF - Own work. Licensed under CC BY-SA 3.0 via Commons.

429. See Chapter 6.

430. One of Hilbert's goals described in Chapter 6 was solution of the *decision problem*, which essentially demands a procedure for deciding which mathematical propositions are true. Turing was not first to prove that such a procedure cannot exist. But he supplied a simple proof using the Turing machine. For details about the history of proofs of this important result, see Wikipedia "Turing machine."

431. The section is based on [Zuse, 1993]; on the material of `http://www.horst-zuse.homepage.t-online.de/` maintained by Horst Zuse, Konrad Zuse's son; and on Wikipedia "Konrad Zuse."

432. Source: "Konrad Zuse" supplied by copyright holder Horst Zuse, who kindly gave permission to use the photo.

433. Source: "Z1 im Wohnzimmer der Eltern (1938), Berlin-Kreuzberg, Wrangelstraße 37" photo supplied by copyright holder Horst Zuse, who kindly gave permission to use the photo.

434. In the floating point representation, a real number is recorded by a *significand*, which has the significant digits, and an *exponent*, which determines a scaling of the significand using some fixed *base*, typically 2, 10, or 16. The formula is $number = significand \cdot base^{exponent}$. See Wikipedia "Floating point" for details.

435. p. 38, 39 [Zuse, 1993].

436. Source: "Reconstruction of Z1, 1989" supplied by copyright holder Horst Zuse, who kindly gave permission to use the photo.

437. Source: "Drawing of Z2" supplied by copyright holder Horst Zuse, who kindly gave permission to use the drawing.

438. Source: "Nachbau der Maschine Z3 (ca. 1961)" supplied by copyright holder Horst Zuse, who kindly gave permission to use the photo.

439. Recall that a computer is *Turing-complete* if it can compute everything computable by the Turing machine, and that all modern computers are Turing-complete. Accordingly, the claim that the Z3 is the first programmable, fully automatic computer can be made only if it is Turing-complete. It is not so obvious that the Z3 has this feature since its program tape does not allow it to branch to different parts of the input program depending on values that have been

computed so far. But the question about Turing-completeness of the Z3 was finally settled in 1998 in the affirmative, and thus there is no doubt whatsoever that the Z3 is indeed the world's first programmable, fully automatic computer. For details, see Wikipedia "Z3 (computer)."

440. The Z4 can branch to different parts of the input program depending on values that have been computed so far, and is obviously Turing-complete.

441. Source: https://upload.wikimedia.org/wikipedia/commons/b/bc/Zuse-Z4-Totale_deutsches-museum.jpg. "Z4 on display at the Deutsches Museum, Munich" by Clemens PFEIFFER - CANON PowerShot G7. Licensed under CC BY 2.5 via Commons.

442. The Swiss Federal Institute of Technology charged 0.01 Suisse Franc per executed instruction for outside use of the Z4 computer. The Z4 processed about 1,000 instructions per hour. For details, see http://www.horst-zuse.homepage.t-online.de/Konrad_Zuse_index_english_html/rechner_z4.html/ Accounting for inflation and exchange rates, in 2016 the rate would have been 0.04 US dollars per instruction, or 40 US dollars per hour. A very fast computer of 2016 processed more than $2 \cdot 10^{11}$ instructions per second; see Wikipedia "Instructions per second" for the performance of various machines since 1951. If such a machine had charged the rate per instruction applied to the Z4 computer, it would have earned the Gross National Product of the US for 2016, which was between 16 and 17 trillion (short scale) US dollars—see http://www.tradingeconomics.com/united-states/gross-national-product—in less than 4 minutes.

443. Search Internet for "Horst Zuse homepage."

444. For details about these developments, see Wikipedia "Computer."

Chapter 8 Question: Creation or Discovery?

445. In 2017, the Internet query "mathematics creation or discovery" produced 2 million answers.

446. See Wikipedia "Archimedes' screw."

447. See Wikipedia "Constitutional Convention (United States)."

448. See Wikipedia "History of Antarctica."

449. See Wikipedia "Nucleic acid double helix."

450. Source: `https://en.wikipedia.org/wiki/Archimedes%27_sc rew#/media/File:IMG_1729_Gemaal_met_schroef_van_Archimedes _bij_Kinderdijk.JPG`. "Archimedes screw" by Ellywa (assumed, based on copyright claims). Own work assumed (based on copyright claims). Licensed under CC BY-SA 2.5 via Commons. For an animated drawing that explains the fundamental idea behind the Archimedean screw, see Wikipedia "Archimedes' screw."

451. Source: `https://en.wikipedia.org/wiki/DNA#/media/File: DNA_orbit_animated_static_thumb.png`. "Section of DNA" by Zephyris. Licensed under CC BY-SA 3.0 via Commons.

452. Search Internet using "create definition" and "discovery definition."

453. The impressionists sometimes used a bluish tint to depict shade, in contrast to the traditional method where the actual color is darkened. Amazingly, the human brain interprets bluish patches as shaded areas.

454. The example demonstrates that deciding the meaning of a sentence when one uses just the sense definitions of the individual words plus their grammatical relationships, can be rather difficult; indeed, it generally is not possible. Chapter 9 addresses the difficulty of determining the meaning of sentences, and the confusion one may encounter in the process.

455. Source: `https://en.wikipedia.org/wiki/Impressionism# /media/File:Claude_Monet,_Impression,_soleil_levant.jpg`. "Impression, soleil levant (Impression, Sunrise)" by Claude Monet - wartburg.edu. Licensed under Public Domain via Commons.
The painting became the source of the label *impressionism* when art critic Louis Leroy published a satirical article titled "The Exhibition of the Impressionists" that declared Monet's painting to be at best a sketch. Leroy didn't realize that he had witnessed the birth of one the most influential art movements in history.

456. p. 5 [Dedekind, 1872].

457. p. 57 [Weyl, 1921], reprinted p. 175 [Thiel, 1982].

458. In a private communication, Christian Thiel of the University of Erlangen/Nürnberg pointed out this fact, adding that Dedekind

and Weyl are "hedging their bets."

459. For details, see Wikipedia "Equivalence class."

460. p. 39 [Whitehead, 1978]. See also Wikipedia "Process and Reality."

461. Source: `https://en.wikipedia.org/wiki/Plato#/media/F ile:Plato_Silanion_Musei_Capitolini_MC1377.jpg`. By English: Copy of Silanion - Marie-Lan Nguyen (User:Jastrow) 2009. Licensed under CC BY 2.5 via Commons.

462. See Wikipedia "Platonism."

463. Quoted by Euthydemos 290 BCE, translation by C. Thiel; personal communication.

464. Our presentation about mathematical platonism is greatly simplified; for a nuanced discussion, see Wikipedia "Platonism" and Stanford Encyclopedia of Philosophy "Platonism in the Philosophy of Mathematics."

465. For other versions, see Stanford Encyclopedia of Philosophy "Platonism in the Philosophy of Mathematics."

466. See Wikiquote "Carl Friedrich Gauss," letter to Bessel, 1830.

467. See Wikiquote "Georg Cantor."

468. p. 13 [Dedekind, 1872].

469. p. 19 [Dedekind, 1872]. The construction process is now called the Dedekind cut.

470. See Wikipedia "Leopold Kronecker."

471. p. 99 [Frege, 1884].

472. See Stanford Encyclopedia of Philosophy "Kurt Gödel."

473. Source: `https://en.wikipedia.org/wiki/Roger_Penrose# /media/File:Roger_Penrose-6Nov2005.jpg`. "Roger Penrose" by Festival della Scienza. Licensed under CC BY-SA 2.0 via Commons.

474. p. 17 [Penrose, 2004].

475. p. 18 [Penrose, 2004].

476. Source: "Armand Borel" Herman Landshoff photographer. From

the Shelby White and Leon Levy Archives Center, Institute for Advanced Study, Princeton, NJ, USA. The institute kindly granted permission to use the photo.

477. [Borel, 2017].

478. See Stanford Encyclopedia of Philosophy "Platonism in the Philosophy of Mathematics."

Chapter 9 Wittgenstein's Philosophy

479. See [Monk, 1990] and Wikipedia "Ludwig Wittgenstein" for details of Wittgenstein's life.

480. Source: https://en.wikipedia.org/wiki/Ludwig_Wittgenstein#/media/File:Ludwig_Wittgenstein.jpg. "Ludwig Wittgenstein" by Moritz Nähr - Austrian National Library. Public Domain under US copyright code PD-old-70.

481. The fly bottle is a passive device for trapping flies. They enter the bottle in search of food and then are unable to escape due to the color or shape of the bottle. The European version is conical and has small feet that raise the bottle about half an inch from the supporting surface.

(Source: https://commons.wikimedia.org/wiki/File:3_fly-bottles.jpg. "3 Fly Bottles" by Sebasstian - Own work. Licensed under CC BY-SA 2.0 via Commons.)
The bottom has a central opening with a raised boundary that forms a circular trough with the wall of the bottle. The trough is filled with beer or vinegar. The bottle is placed on a plate on which sugar has been sprinkled to attract the flies. After feeding, the fly rises up and enters the bottle. Once inside, the fly moves toward the light coming through the wall of the bottle and doesn't try to escape through the hole in the center. Eventually, the fly falls into the liquid in the trough and drowns. For details and variants of the fly bottle, see Wikipedia "Fly-killing device."

482. Paragraph 309 [Wittgenstein, 1958]. See also Stanford Encyclopedia of Philosophy "Ludwig Wittgenstein."

483. We emphasize that the definition of *valid* introduced here in the context of philosophical questions and statements differs from that in the earlier discussion of mathematical concepts. There, an axiom or theorem is declared to be *valid* if it is true.

484. We ignore here the role of contradictions and tautologies since they are not essential for the arguments of this book.

485. We were taught in high school that the answer undoubtedly is "no."

486. [Kahneman, 2011] provides a nuanced discussion of this process. The author differentiates between fast thinking, which is done initially, without conscious awareness and seemingly instantaneously, and slow thinking, which is deliberate and done consciously.

487. See Wikipedia "Language-game (philosophy)."

488. Paragraph 1 [Wittgenstein, 1958]. The cited statement is based on a quote of Augustinus who summarized how he learned language. Essentially Augustinus observed how adults labeled things using words, and from these observations he deduced the meaning of words for objects.

489. Paragraph 2 [Wittgenstein, 1958].

490. Paragraph 6, 7 [Wittgenstein, 1958].

491. Source: https://en.wikipedia.org/wiki/Johann_Wolfgang_von_Goethe#/media/File:Goethe_(Stieler_1828).jpg. "Johann Wolfgang Goethe" by Joseph Karl Stieler, 1828, detail. Transferred from nds.wikipedia to Commons by G. Meiners 2005. Public Domain.

492. See Wikipedia "Ludwig Wittgenstein" for an overview. For some details, go to Wikipedia "On Certainty" about discussion of doubt; "Remarks on Colour" for analysis of Goethe's theory of color; "Private language argument" for investigation of the existence of a private language; "Zettel (Wittgenstein)" for discussion of philosophical psychology; and Stanford Encyclopedia of Philosophy "Wittgenstein's Philosophy of Mathematics" for wide ranging treatment of mathematics.

493. p. viii [Wittgenstein, 1958].

494. [Wittgenstein, 1963].

495. The Tractatus consists of declarative numbered sentences that make various claims about the world, language, the link between the world and language, and the limits of language. For example, the first two statements are "1. The world is all that is the case" and "1.1 The world is the totality of facts, not of things."
The terse statements of the Tractatus are difficult to understand. But there is extensive help available, in the form of a number of introductory texts, for example, [Anscombe, 1971], [Hartnack, 1965], and [Fann, 2015]. The latter two books are particularly easy to follow. An introductory text that illustrates Wittgenstein's arguments using concepts of physics, is [Hülster, 2017].
It's beyond the scope of this note even to outline the main thoughts and arguments of the Tractatus. But we summarize two key ideas that now are called the *picture theory* of the Tractatus and *logical atomism*.
The *picture theory* postulates that each sentence of language is a picture of part of the world, in the sense that the logic relationships tying the meanings of the words together must correspond to relationships among the corresponding things of the world. Wittgenstein first had this idea when he observed how a traffic accident was reenacted in a traffic court in Paris: Toy cars were moved by hand on a scaled model of streets and intersections, just as real cars had moved. He then considered this reenactment with a physical model to be analogous to language depicting situations of the world.
Logical atomism postulates that the world consists of ultimate facts that cannot be further subdivided or broken down. These atomic facts are combined using mathematical logic to create objects of the world. When we invert this process, we begin with objects of the world. Each object is then broken down into constituent parts, which in turn can be broken down further, and so on. Eventually, this process must stop; the ultimate constituents are called facts. For example, a car consists of a body and wheels. Each wheel is made up of a tire and a rim. The tire consists of rubber particles, and so on. The process stops when no further breakdown is possible.
Wittgenstein argued that all declarative statements of a language are consistent with his picture theory, with meaning of individual words defined via logical atomism, and that any declarative statement that is not of that form is nonsensical. In particular, most philosophical propositions are then nonsensical. For example, the question "What is the value of the soul?" and the claim "The out-

side world is not real; only my perceptions of it are real" are non-sensical.

In the second half of the 1920s, Wittgenstein realized that the postulates of logical atomism are not valid, and that his picture theory failed to capture the richness of human language and its use. But he still insisted that philosophical statements are largely mistaken, and set out to find a new way to cope with such statements. He came up with the approach mentioned at the beginning of the chapter: the language game.

496. See above note.

497. [Wittgenstein, 1958].

498. See Wikipedia "Ludwig Wittgenstein."

499. See above note.

500. As done earlier, we ignore the role of tautologies and contradictions.

501. p. 108 [Hülster, 2017] restates this as, "Forget these propositions [of the Tractatus] as quickly as possible, and then try again to go back to making philosophical statements—if you can. It can no longer be done."

502. Source: https://en.wikipedia.org/wiki/Ludwig_Wittgenstein#/media/File:Wittgenstein_Gravestone.jpg. "Wittgenstein gravestone" photo by Andrew Dunn. Licensed under CC BY-SA 2.0 via Commons.
From Wikipedia "Ludwig Wittgenstein:"
"Ludwig Wittgenstein's gravestone in the graveyard of the chapel for Ascension Parish Burial Ground off Huntingdon Road, Cambridge, UK. The graveyard was previously known as St. Giles Cemetery in association with the parish of St. Giles church at the bottom of Castle Hill ...
"The grave is usually lightly covered with large pine needles from the overlooking trees. It has a small but steady stream of visitors, as implied by the fresh carnation somebody has left ... [The ladder] is part of a ... trend to leave small items at the grave, as are the scattering of pennies and the votive candles."

503. [Carnap, 1928] treats this problem and declares the cited statement to be nonsensical. His criterion for valid statements is "Sachhaltigkeit" (objective relevance). The statement fails that test.

504. p. viii [Wittgenstein, 1958].

505. [Goethe, 1810].

506. [Wittgenstein, 1978].

507. Chapter "Entwurf einer Farbenlehre Didaktischer Teil" of [Goethe, 1810].

508. Part I, paragraph 63 [Wittgenstein, 1978].

509. Source: https://en.wikipedia.org/wiki/Chess_piece#/media/File:RosewoodPieces.jpg. "Staunton pieces made of rosewood" by Bubba73 at en.wikipedia. Licensed under CC BY-SA 3.0 via Commons.

510. Source: https://en.wikipedia.org/wiki/Claude_Monet#/media//File:Claude_Monet_-_Woman_with_a_Parasol_-_Madame_Monet_and_Her_Son_-_Google_Art_Project.jpg. "Woman with a Parasol 1875" by Claude Monet - EwHxeymQQnprMg at Google Cultural Institute maximum zoom level. Licensed under Public Domain via Commons.

511. The arguments are based on paragraph 2 [Wittgenstein, 1978]: "In a picture in which a piece of white paper gets its lightness from the blue sky, the sky is lighter than the white paper. And yet in another sense blue is the darker and white the lighter colour (Goethe). On the palette white is the lightest colour."

512. See Wikiquote "Claude Monet" for the quote of Paul Cézanne.

513. [Wittgenstein, 1956].

514. See Wikipedia "Remarks on the Foundations of Mathematics."

515. See Stanford Encyclopedia of Philosophy "Wittgenstein's Philosophy of Mathematics."

Chapter 10 Language Games of History

516. [Hersh, 2014] uses delightful dialogues to clarify certain philosophical claims. In Wittgenstein's terminology, the dialogues are language games.

517. Chapter 2 [Rudman, 2007].

518. p. 10 [Dedekind, 1872].

519. See Wikiquote "Carl Friedrich Gauss," letter to Bessel, 1830.

520. See Chapter 3.

521. Leibniz published his results in three papers during the period 1684–1693: *Nova Methodus pro Maximis et Minimis* ... (1684), *De geometria recondite et analysi indivisibilium atque infinitorum* ... (1686), and *Supplementum Geometrie Dimensorie* ... (1693). For details about the papers, go to http://www.maa.org/book/export/html/641727. Newton began work in 1665 and published the results in the paper *Method of Fluxions* in 1671. For details, see pp. 192–198 [Cajori, 1919b].

522. See Chapter 3.

523. Folio 249 verso [Stifel, 1544].

524. Source: Toggenburger Museum, Lichtensteig, Switzerland. The museum kindly has granted permission to use the photo.

525. Since the numbers of the outer ring of the title page, which are logarithms of the numbers of the inner ring, are uniformly increasing around the circle, they effectively correspond to uniform changes of angles. So if the outer ring of the displayed circular slide rule is rotated relative to the inner ring, then effectively logarithm values of the displayed numbers are added or subtracted. Wikipedia "Slide rule" includes modern circular slide rules and their operation.

526. See Wikipedia "William Oughtred."

527. Source: "Circular Slide Rule" assembled by Ingrid Truemper from the photo supplied by the Toggenburger Museum; see above note. Inset: slide rule with plastic disks by Klaus Truemper, photo released into Public Domain under Creative Commons CC0.

528. Source: https://en.wikipedia.org/wiki/William_Oughtred#/media/File:Wenceslas_Hollar_-_William_Oughtred.jpg. "William Oughtred" by Wenceslaus Hollar - Artwork from University of Toronto Wenceslaus Hollar Digital Collection. Scanned by University of Toronto. High-resolution version extracted using custom tool by User:Dcoetzee. Licensed under Public Domain via Commons.

529. For an expanded discussion of this input, see Chapter 6 about the well-ordering theorem.

530. See Chapter 6.

531. See Wikipedia "Well-order."

532. See Chapter 3.

533. Source: `https://en.wikipedia.org/wiki/Lebesgue_integ ration#/media/File:RandLintegrals.png`. "RandLintegrals" by en.wiki. Licensed under CC BY-SA 3.0 via Commons.

534. See Chapter 3.

535. See Chapter 4.

536. This conjecture is the continuum hypothesis. Cantor indeed believed it to be true, but Gödel and Cohen proved that it is independent from ZF as well as ZFC. See Chapter 6.

537. See Wikipedia "Fermat Number."

538. See Wikipedia "Leonhard Euler" for related results of Euler in number theory.

539. Statement A p. 493 [Cantor, 1895].

540. The axiom of countable choice assures that every infinite set has a countably infinite subset. See Wikipedia "Axiom of countable choice."

541. See Frege's and Gödel's statements in Chapter 8, which are general versions of this argument.

542. A notable exception is Gauss, who said, "I confess that Fermat's Theorem as an isolated proposition has very little interest for me, because I could easily lay down a multitude of such propositions, which one could neither prove nor dispose of." See Wikiquote "Carl Friedrich Gauss." It is a rare exception to his generally farsighted observations and comments.

543. [Wigner, 1960] and [Hamming, 1980].

544. [Abbott, 2013].

545. See Wikipedia "Albert Einstein."

546. See Chapter 3 for details about Cartesian coordinate systems.

547. Source: `https://commons.wikimedia.org/wiki/File:Migu el_%C3%81ngel,_por_Daniele_da_Volterra_(detalle).jpg`.

"Michelangelo" by Daniele da Volterra - The Collection Online, The Metropolitan Museum of Art. Licensed under Public Domain via Commons.

548. Source: https://en.wikipedia.org/wiki/Michelangelo#/media/File:%27David%27_by_Michelangelo_JBU0001.JPG. "David by Michelangelo" by Jörg Bittner Unna - Own work. Licensed under CC BY 3.0 via Commons.

549. Gödel ingeniously constructed within the given framework a self-referential statement that effectively says, "I am not provable." It is then easy to see that the statement is true using elementary logic. See Wikipedia "Proof sketch for Gödel's first incompleteness theorem" and Chapter 6.

550. Source: https://en.wikipedia.org/wiki/Imre_Lakatos#/media/File:Professor_Imre_Lakatos,_c1960s.jpg. "Imre Lakatos" by Library of the London School of Economics and Political Science, circa 1960. No restrictions, see https://commons.wikimedia.org/w/index.php?curid=15336126.

551. [Lakatos, 2015]

552. Source: https://en.wikipedia.org/wiki/Polyhedron#/media/File:Icosidodecahedron. "Icosidodecahedron" by Tomruen - Own work, Copyrighted free use.

553. Source: https://en.wikipedia.org/wiki/Stellated_octahedron#/media/File:Stellated_octahedron_3-fold-axis.png. "Stellated Octahedron" by Tomruen - Own work. Licensed under Creative Commons CC BY-SA 4.0.

554. Source: https://upload.wikimedia.org/wikipedia/commons/4/4b/Hexagonal_torus.png. "Hexagonal torus" by Tom Ruen (talk). Public Domain.

555. The discussion of this section relies entirely on [Lakatos, 2015]. The book contains a wealth of historical information and references concerning Euler's formula. We omit almost all of that detail and refer to reader to the book for all references.

556. The mathematician Reuben Hersh interpreted the mathematical process in similar fashion; see [Hersh, 1995]. The following paraphrased version by the mathematician Jonathan Borwein is quoted in [Théra, 2017]:
"Mathematics is a human endeavor. It takes part in and adapts it-

self to culture. It is not a question of abstract reality, immutable, eternal, unearthly constructs as conceived of by Frege.

"Mathematical knowledge is not infallible. In concert with empirical science, mathematics can move forward while errors are made, then corrected, and then perhaps corrected again. This flawed nature of the subject is brilliantly described in 'Proofs and Refutations' by Lakatos.

"There exist several conceptions of proof and of mathematical rigor, as a function of time, place and other considerations. The use of computers to construct proofs constitutes a nontraditional version of rigor.

"Empirical evidence, numerical techniques, and probabilistic evidence help all of us to decide what ought to be believed as true in mathematics. Aristotelean logic isn't always the best means to come to a decision."

557. See Chapter 9.

558. Source: https://en.wikipedia.org/wiki/Johann_Sebastian_Bach#/media/File:Johann_Sebastian_Bach.jpg. "Johann Sebastian Bach" (aged 61) by Elias Gottlob Haussmann - http://www.jsbach.net/bass/elements/bach-hausmann.jpg. This is a copy or second version of his 1746 canvas. The original painting hangs in the upstairs gallery of the Altes Rathaus (Old Town Hall) in Leipzig, Germany. Licensed under Public Domain via Commons.

559. Source: https://en.wikipedia.org/wiki/Wolfgang_Amadeus_Mozart#/media/File:Wolfgang-amadeus-mozart_1.jpg. "Wolfgang Amadeus Mozart." Posthumous painting by Barbara Krafft, 1819. Licensed under Public Domain via Commons.

560. See Chapter 6 and the discussion about knowledge of function values earlier in this chapter.

561. Divertimento for two horns and string quartet. Detailed notes about the piece and a complete performance are available at https://www.youtube.com/watch?v=wzaQixVGoQg.

562. See Chapter 6.

563. See Chapter 4.

564. See Chapter 6.

565. See the discussion earlier in this chapter and Stanford Ency-

clopedia of Philosophy "Zermelo's Axiomatization of Set Theory."

566. See Stanford's Encyclopedia of Philosophy "Hilbert's Program."

567. Source: https://upload.wikimedia.org/wikipedia/commons/8/8c/Title_page_William_Shakespeare%27s_First_Folio_1623.jpg. "William Shakespeare" by Martin Droeshout. Part of the title page of the first folio of *Comedies, Histories, and Tragedies*, 1623.

568. Source: https://en.wikipedia.org/wiki/Romeo_and_Juliet#/media/File:Francesco_Hayez_053.jpg. "L'ultimo bacio dato a Giulietta da Romeo" by Francesco Hayez - The Yorck Project: 10.000 Meisterwerke der Malerei. DVD-ROM, 2002. ISBN 3936122202. Distributed by DIRECTMEDIA Publishing GmbH. Licensed under Public Domain via Commons.

569. See Wikipedia "Ancient Greek sculpture."

Chapter 11 Effectiveness of Mathematics

570. See Wikipedia "History of science."

571. See Wikipedia "Modern physics."

572. Examples are the books [Hawking and Mlodinow, 2010] and [Penrose, 2004].

573. [Hawking and Mlodinow, 2010].

574. Source: https://en.wikipedia.org/wiki/Global_Positioning_System#/media/File:GPS_Satellite_NASA_art-iif.jpg. "Artist's conception of GPS Block II-F satellite in Earth orbit" by NASA. Licensed under Public Domain via Commons.

575. See Wikipedia "Global Positioning System."

576. [Wigner, 1960].

577. [Hamming, 1980].

578. p. 90 [Hamming, 1980] says: "From all this, I am forced to conclude both that mathematics is unreasonably effective and that all the explanations I have given when added together simply are not enough to explain what I set out to account for."

579. [Abbott, 2013].

580. We shall not cover the three papers in detail, but encourage the reader to do so. The first two papers on the unreasonable effectiveness of mathematics offer an opportunity to apply Wittgenstein's method of language games for understanding where and how arguments get derailed. This is not meant as an arrogant comment. The matter of fact is that philosophical statements almost always involve errors that are virtually impossible to detect. Wittgenstein's method helps in that detection process. The third paper avoids this trap of philosophical statements by looking at the issue in various ways, just as Wittgenstein recommends.

581. See Wikipedia "Antikythera mechanism."

582. See Wikipedia "Ptolemy."

583. See Wikipedia "Apparent retrograde motion."

584. Source: https://en.wikipedia.org/wiki/Ptolemy#/media/File:PSM_V78_D326_Ptolemy.png. "Claudius Ptolemy" by Unknown - reproduced from Popular Science Monthly Volume 78, April, 1911, p. 316. Licensed under Public Domain via Commons.

585. For a complete explanation of the X and the large dot, see Wikipedia "Deferent and epicycle." Source: https://en.wikipedia.org/wiki/Deferent_and_epicycle#/media/File:Ptolemaic_elements.svg. "Deferent and epicycle" by Fastfission. Licensed under Public Domain via Commons.

586. See Wikipedia "Nicolaus Copernicus."

587. Source: https://en.wikipedia.org/wiki/Johannes_Kepler#/media/File:Johannes_Kepler_1610.jpg. "Johannes Kepler" by Unknown. Licensed under Public Domain via Commons.

588. The five platonic solids: tetrahedron, cube, octahedron, dodecahedron, and icosahedron, with 4, 6, 8, 12, and 20 facets:

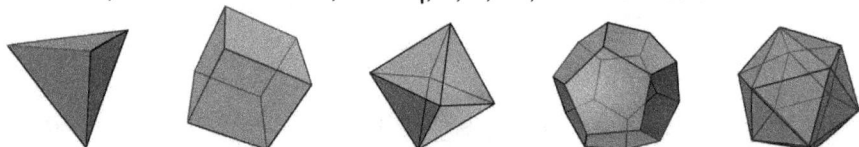

(Source: https://en.wikipedia.org/wiki/Platonic_solid. Vectorization of images by DTR and Stannered. Licensed under CC BY-SA 3.0 via Commons.) For details about the platonic solids, see Wikipedia "Platonic solids."

589. See Wikipedia "Johannes Kepler."

590. Source: https://en.wikipedia.org/wiki/Johannes_Kepler#/media/File:Kepler-solar-system-1.png. Diagram of Platonic solids model of the solar system taken from *Mysterium Cosmographicum*, 1596. Licensed under Public Domain via Commons.

591. Source: "Ellipse with sun and planet" by K. Truemper, released into Public Domain under Creative Commons CC0.

592. See Wikipedia "Ellipse."

593. See Wikipedia "Isaac Newton."

594. See Wikipedia "*n*-body problem."

595. See Wikipedia "Three-body problem."

596. See Wikipedia "General relativity."

597. See Wikipedia "Solar system."

598. See Wikipedia "Solar system."

599. See Wikipedia "Chaos theory."

600. See Wikipedia "David Orrell."

601. See Wikipedia "Weather forecasting."

602. See Wikipedia "Nowcasting (meteorology)."

603. Source: https://en.wikipedia.org/wiki/Weather_forecasting#/media/File:2005-09-22-10PM_CDT_Hurricane_Rita_3_day_path.png. "Hurricane Rita path prediction" by NOAA. Licensed under Public Domain via Commons.

604. See Wikipedia "Economic model."

605. See Wikipedia "Economic forecasting."

606. See Wikipedia "Economic model."

607. See Wikipedia "David Orrell."

608. See Wikipedia "Computational economics."

609. See Wikipedia "Nowcasting (economics)."

Chapter 12 Life Without Mathematics

610. pp. 122–126 [Everett, 2008] describe a typical situation.

611. Source: http://languagelog.ldc.upenn.edu/myl/ldc/llog /Everett_map1a.gif. "Brazil map," copyright Daniel Everett, who kindly gave permission for use of the drawing.

612. The material of the next two sections is based on [Everett, 2008] and [Everett, 2012]. For an introduction, see [Colapinto, 2007].

613. See Wikipedia "Cetacean intelligence."

614. Source: "Daniel Everett" by KristenN2013 - Own work. High resolution version provided by Daniel Everett, who kindly granted permission to use the photo.

615. Source: https://en.wikipedia.org/wiki/Noam_Chomsky#/me dia/File:Noam_Chomsky_portrait_2015.jpg. "Noam Chomsky" by https://www.flickr.com/photos/culturaargentina. The file has been extracted from another file: Noam Chomsky.jpg. Licensed under CC BY-SA 2.0 via Commons.

616. See Wikipedia "Noam Chomsky."

617. The sentences of the panther story on pp. 123–126 [Everett, 2008] are one such example.

618. Paragraph 19 [Wittgenstein, 1958].

619. [Everett, 2012].

620. See Wikipedia "For sale: baby shoes, never worn."

621. See the discussion by the Quote Investigator http://quoteinv estigator.com/2014/06/14/exclamation/ according to which the claimed exchange of telegrams very likely did not occur.

622. See Wikipedia "Noam Chomsky" and "Transformational grammar."

623. The Google Translator translates "I love you, honey" to German "Ich liebe Dich, Honig," which is nonsense since "Honig" always refers to the product of bees. The Babylon and SDL translators deliver "Ich liebe sie, Honig," which is worse since the formal address of "Sie" is used, here misspelled in lower case, instead of the implied informal "Dich." See also the introductory section of

Chapter 13.

624. This section is based on pp. 116–119 [Everett, 2008] and pp. 259–262 [Everett, 2012].

625. p. 261, 262 [Everett, 2012].

Chapter 13 Brain Science

626. See Wikipedia "Big Bang."

627. See Chapter 9.

628. See Chapter 9.

629. [Møller, 2015] has a detailed survey.

630. In a much more sophisticated and extensive use of brain science, [Lakoff and Núñez, 2000] employs the concept of *embodied cognition* to explain where mathematics comes from. The notion of embodied cognition subsumes that of embodied simulation invoked later in this chapter.

The key cognitive concept of the book is *metaphor*. In general, a metaphor applies insight in one area to another one. In the book, the metaphors take everyday intuitive arguments into the world of mathematics, thus producing mathematical insight. For example, the *Basic Metaphor of Infinity* uses an intuitive understanding of infinite processes that achieve a final state, to arrive at conclusions about various infinite mathematical processes.

The book also addresses the question of creation versus discovery of mathematics. It declares that mathematics is created, just as we conclude later. The book does not offer a proof, indeed says there cannot be a proof, something we agree with. But this still leaves the door open for the approach taken here, where validity of the creation claim is *demonstrated* in the sense of Wittgenstein.

631. [Wittgenstein, 1963].

632. See Chapter 9.

633. See Chapter 9.

634. For decades, the Artificial Intelligence community ignored Wittgenstein's results, insisting on computation of meaning of sentences via methods that effectively used the picture theory. As would have been predicted by Wittgenstein, this approach fared badly, as is evi-

dent from the failure of language translation over decades of effort. For a description of efforts that effectively are based on Wittgenstein's picture theory, see pp. 907–912 [Russell and Norvig, 2010].

635. We cannot possibly do justice to the various research results and arguments for or against various concepts. For example, see Wikipedia "Speech perception" and "Visual perception."

636. See Merriam Webster Dictionary http://www.merriam-webster.com/dictionary/embodiment.

637. An important feature of the brain makes embodiment of experiences and the later described rerunning of experiences in embodied simulation possible: the *neuroplasticity* of the brain. It is yet another stunning concept of modern brain science. Generally speaking, it says that the brain continuously modifies itself regardless of age. See [Møller, 2014] for an in-depth treatment and Wikipedia "Neuroplasticity" for a general discussion.

638. [Zeki et al., 2014] shows that a certain area of the brain is involved in the evaluation of beauty, be it mathematical, visual, musical, or even moral.

639. [Coello and Fischer, 2016] and [Fischer and Coello, 2015] are the first two volumes of a planned four-volume set that covers the results of a large number of research studies concerning embodiment and *embodied cognition*. The latter area is concerned with the brain's running of experiences and evaluation of the results. [Bergen, 2012] provides an introduction to embodied simulation for the nonspecialist; the case of movies is covered in Chapter 2 of [Zacks, 2015].

640. [Gallese and Sinigaglia, 2011] provides a detailed discussion of the various arguments and concludes that the claims of embodied simulation are well-justified.
We emphasize that our discussion of embodied simulation is just a sketch of the results and omits a large body of related material such as the concept of *Analysis by Synthesis*; for a review, see [Bever and Poeppel, 2010]. It is impossible for us to include details in this book. It also is not necessary, since the main goal of this chapter isn't a nuanced presentation of embodied simulation and related concepts of brain science. Instead, we just want to show that important brain science results lead to a conjecture about the causes of philosophical confusion.

641. p. x of foreword by George Lakoff in [Bergen, 2012].

642. [Everett, 2008] and [Everett, 2012].

643. For example, [Moseley et al., 2016] describes determination of the concrete meaning of "eye" and of the abstract meaning of "beauty," [Garagnani and Pulvermüller, 2013] deals with the decision to speak or act, and [Pulvermüller et al., 2013] proposes a model for representation of syntax and grammar.

644. For example, embodied simulation so far has not been linked to the fast and slow thinking processes of [Kahneman, 2011].

645. Since current knowledge about the brain is incomplete, it may well be, indeed is likely, that the concepts of embodiment and embodied simulation will be modified and adapted. But we believe that the main conclusion used here, which says that the brain relies on learning experiences to make decisions about the world, will remain intact.

646. [Kahneman, 2011] separates decision processes into fast and slow thinking. In the terminology used here, fast thinking consists of embodied simulation plus an evaluation that is carried out at extraordinary speed and without conscious awareness; indeed, fast thinking is not open to direct investigation. In contrast, slow thinking includes deliberate subsequent logic reasoning. We simplify the discussion here to bring out the role of embodied simulation versus that of logic reasoning.

647. See Wikipedia "Carl Friedrich Gauss."

648. See Wikiquote "Carl Friedrich Gauss," letter to Bessel, 1830.

649. p. 99 [Frege, 1884].

650. See Stanford Encyclopedia of Philosophy "Kurt Gödel."

651. See Chapter 6.

652. We once heard in a mathematics class, "The axioms are the foundation of mathematics. Below them is quicksand."

653. In 2013, we gave a talk about creation and discovery of mathematics at a workshop. We thought that we had clearly outlined the case for creation. But after the talk, a number of the 50-60 participants asked us what our opinion was—which surprised us—and almost all of them argued strongly for discovery.
An explanation would be that the information of the talk was evaluated by the brains of the listeners using earlier learning about

mathematics. Embodied simulation then typically led to the conclusion that, first, the talk didn't reveal the opinion of the presenter, and second, that mathematics was discovered. The talk is available at http://www.utdallas.edu/~klaus/Wpapers/math-discovered-constructed.pdf.

654. See Wikipedia "Experimental mathematics."

655. See Note 556, where Jonathan Borwein paraphrases a formulation of the mathematical process by Reuben Hersh.
Borel argues in [Borel, 2017]: "... [C]omputers have been playing an increasing role [in experiments]. They have given this experimental side of mathematics a new dimension. This has advanced to the extent that one can already see important, reciprocal and fascinating interactions between computer science and pure mathematics."

656. [Abbott, 2013] mentions informal surveys where, in the terminology used here, engineers voted largely for creation, physicists in public tended to support discovery but secretly favored creation, and mathematicians largely decided for discovery.

657. See Wikiquote "Carl Friedrich Gauss," Gauss-Schumacher Briefwechsel, 1862.

658. See opening paragraph in Wikipedia "Quality (philosophy)."

659. See Wikipedia "Rasmus Sørnes" for details about this gifted inventor and clock maker and his extraordinary Clock Number 4.

660. Source: "Sørnes Clock Number 4" by photographer and copyright holder Stephen Pitkin, who kindly gave permission to use the photo. The clock has height 210cm (82.6in), depth 60cm (23.6in), and width 70cm (27.6in). For technical and historical details about the clock and the three predecessor clocks designed and built by R. Sørnes, see [Sørnes, 2008].

661. Source: https://en.wikipedia.org/wiki/Canaletto#/media/File:Piazza_San_Marco_with_the_Basilica,_by_Canaletto,_1730._Fogg_Art_Museum,_Cambridge.jpg. "Piazza San Marco, Venice" by Giovanni Antonio Canal, better known as Canaletto, ca. 1730–1735, detail. Fogg Museum, Cambridge, Massachusetts. Licensed under Public Domain via Commons.

Chapter 14 Conclusion: Creation

662. See Wikipedia "Occam's razor."

663. Some relevant language games are already included in Chapter 10. See also [Théra, 2017] for a response based on the power of computing.

664. See Chapter 13.

665. See Wikipedia "Mathematical beauty."

666. See Chapter 13.

Bibliography

[Abbott, 2013] Abbott, D. (2013). The reasonable ineffectiveness of mathematics. *Proceedings of the IEEE*, vol. 101.

[Alexander, 2014] Alexander, A. (2014). *Infinitesimal: How a Dangerous Mathematical Theory Shaped the Modern World*. Scientific American/Farrar, Straus, and Giroux.

[Anscombe, 1971] Anscombe, G. E. M. (1971). *An Introduction to Wittgenstein's Tractatus*. St. Augustine's Press.

[Archibald, 1920] Archibald, R. C. (1920). Gauss and the Regular Polygon of Seventeen Sides. *American Mathematical Monthly*, vol. 27.

[Badur and Rottstedt, 2004] Badur, K. and Rottstedt, W. (2004). Und sie rechnet doch richtig! Erfahrungen beim Nachbau einer Leibniz Rechenmaschine. *Studia Leibnitiana*, vol. 36.

[Bergen, 2012] Bergen, B. K. (2012). *Louder Than Words: The New Science of How the Mind Makes Meaning*. Basic Books.

[Bever and Poeppel, 2010] Bever, T. G. and Poeppel, D. (2010). Analysis by Synthesis: A (Re-)Emerging Program of Research for Language and Vision. *Biolinguistics*, vol. 4.

[Boole, 1854] Boole, G. (1854). *An Investigation of the Laws of Thought on which are founded the Mathematical Theories of Logic and Probabilities*. Walton & Maberly; go to `https://archive.org/index.php` and search for "George Boole Laws of Thought".

[Borel, 2017] Borel, A. (2017). Mathematics: Art and Science. *European Mathematical Society Newsletter*, March 2017.

[Bos, 1974] Bos, H. J. M. (1974). Differentials, higher-order differentials and the derivative in the Leibnizian calculus. *Archive for History of Exact Sciences*, vol. 14.

[Bourbaki, 1948] Bourbaki, N. (1948). *L'Architecture des Mathématiques*. In *Les grands courants de la pensée mathématique*, Cahiers du Sud, Marseille.

[Bürgi, 1620] Bürgi, J. (1620). *Aritmetische und Geometrische Progress Tabulen*. Paul Sessen, Universitätsbuchdruckerei, Prag; go to ht tp://daten.digitale-sammlungen.de/~db/0008/bsb00082065 /images/index.html?id=00082065&groesser=&fip=193.174.9 8.30&no=&seite=7.

[Cajori, 1918] Cajori, F. (1918). Pierre Laurent Wantzel. *Bull. Amer. Math. Soc.*, vol. 24.

[Cajori, 1919a] Cajori, F. (1919a). *A History of Mathematics*. Second edition, revised and enlarged. Macmillan Company; go to ht tps://archive.org/index.php and search for "A History of Mathematics Florian Cajori".

[Cajori, 1919b] Cajori, F. (1919b). *A History of the Conceptions of Limits and Fluxions in Great Britain*. Open Court Publishing Company; go to https://archive.org/index.php and search for "Limits and Fluxions Florian Cajori".

[Cajori, 1928] Cajori, F. (1928). *A History of Mathematical Notations, Vol. I: Notations in Elementary Mathematics*. Open Court Publishing Company; go to https://archive.org/index.php and search for "Mathematical Notations Florian Cajori".

[Cantor, 1895] Cantor, G. (1895). Beiträge zur Begründung der transfiniten Mengenlehre, Erster Artikel. *Mathematische Annalen*, vol. 46, pp. 481–512.

[Carnap, 1928] Carnap, R. (1928). Scheinprobleme in der Philosophie. reprinted by Meiner Verlag, Hamburg, 2005; go to http://www.blutner.de/philos/Texte/carnap.pdf.

[Coello and Fischer, 2016] Coello, Y. and Fischer, M. H. (2016). *Foundations of Embodied Cognition, Vol. 1: Perceptual and Emotional Embodiment*. Routledge.

[Colapinto, 2007] Colapinto, J. (2007). Has a remote Amazonian tribe upended our understanding of language? *New Yorker*, April 16 issue, 2007.

[Dedekind, 1872] Dedekind, R. (1872). *Stetigkeit und irrationale Zahlen*. Friedrich Bieweg und Sohn; go to https://archive.or g/index.php and search for "Stetigkeit und irrationale Zahlen".

[Dénes, 2011] Dénes, T. (2011). Real Face of János Bolyai. *Notices of the American Mathematical Society*, Jan. 2011.

[Descartes, 1637] Descartes, R. (1637). *Discours de la méthode (Discourse on the Method)*. Ian Maire. Go to http://www.gutenberg. org/ and search for "Discourse of the Method" for the English version.

[Dreyfus, 1965] Dreyfus, S. E. (1965). *Dynamic Programming and the Calculus of Variations*. Academic Press.

[Dunham, 1990] Dunham, W. (1990). *Journey through Genius: The Great Theorems of Mathematics*. Wiley.

[Dunnington, 2004] Dunnington, G. W. (2004). *Carl Friedrich Gauss: Titan of Science*. Mathematical Association of America.

[Engel and Stäckel, 1895] Engel, F. and Stäckel, P. (1895). *Die Theorie der Parallellinien von Euklid bis auf Gauß*. Teubner Verlag; available at https://archive.org/details/theoriederparall 00stac.

[Erdös and Dudley, 1983] Erdös, P. and Dudley, U. (1983). Some remarks and problems in number theory related to the work of Euler. *Mathematics Magazine*, vol. 56.

[Everett, 2008] Everett, D. L. (2008). *Don't Sleep, there are Snakes*. Random House.

[Everett, 2012] Everett, D. L. (2012). *Language – The Cultural Tool*. Pantheon Books.

[Fann, 2015] Fann, K. T. (2015). *Wittgenstein's Conception of Philosophy*. Partridge Publishing.

[Fischer and Coello, 2015] Fischer, M. H. and Coello, Y. (2015). *Foundations of Embodied Cognition, Vol. 2: Conceptual and Interactive Embodiment.* Routledge.

[Frege, 1879] Frege, G. (1879). *Begriffsschrift, eine der arithmetischen nachgebildete Formelsprache des reinen Denkens.* Verlag von Louis Nebert. English translation available at http://dec59.ruk.cuni.cz/~kolmanv/Begriffsschrift.pdf.

[Frege, 1884] Frege, G. (1884). *Die Grundlagen der Arithmetik: Eine logisch mathematische Untersuchung über den Begriff der Zahl.* Verlag von Wilhelm Koebner; go to https://archive.org/index.php and search for "Gottlob Frege Grundlagen der Arithmetik".

[Frege, 1893] Frege, G. (1893). *Grundgesetze der Arithmetik, Begriffsschriftlich abgeleitet.* Verlag von Hermann Pohle; available at http://gallica.bnf.fr/ark:/12148/bpt6k77790t/f3.image.

[Fritz, 1945] Fritz, K. v. (1945). The discovery of incommensurability by Hippasus of Metapontum. *Annals of Mathematics*, vol. 46.

[Gallese and Sinigaglia, 2011] Gallese, V. and Sinigaglia, C. (2011). What is so special about embodied simulation? *Trends in Cognitive Sciences*, vol. 15.

[Garagnani and Pulvermüller, 2013] Garagnani, M. and Pulvermüller, F. (2013). Neuronal correlates of decisions to speak and act: Spontaneous emergence and dynamic topographies in a computational model of frontal and temporal areas. *Brain & Language*, vol. 127.

[Gieswald, 1856] Gieswald, H. R. (1856). *Justus Byrg als Mathematiker, und dessen Einleitung in seine Logarithmen.* St. Johannisschule, Danzig, Prussia; see Google books "justus byrg als mathematiker".

[Goethe, 1810] Goethe, J. W. (1810). *Entwurf einer Farbenlehre.* German and English versions available at https://theoryofcolor.org/Theory+of+Color.

[Grötschel et al., 2016] Grötschel, M., Knobloch, E., Schiffers, J., Woisnitza, M., and Ziegler, G. M., editors (2016). *Leibniz: Vision als Aufgabe.* Berlin-Brandenburgische Akademie der Wissenschaften.

[Hamming, 1980] Hamming, R. W. (1980). The unreasonable effectiveness of mathematics. *The American Mathematical Monthly*, vol. 87.

[Hartnack, 1965] Hartnack, J. (1965). *Wittgenstein and Modern Philosophy.* Methuen.

[Hawking and Mlodinow, 2010] Hawking, S. and Mlodinow, L. (2010). *The Grand Design.* Bantam Books.

[Heath, 1910] Heath, T. L. (1910). *Diophantus of Alexandria: A Study in the History of Greek Algebra.* Cambridge University Press; go to `https://archive.org/index.php` and search for "Heath Diophantus of Alexandria".

[Hersh, 1995] Hersh, R. (1995). Fresh Breezes in the Philosophy of Mathematics. *American Mathematical Monthly*, vol. 102.

[Hersh, 2014] Hersh, R. (2014). *Experiencing mathematics – What do we do, when we do mathematics?* American Mathematical Society.

[Hülster, 2017] Hülster, F. (2017). *Introduction to Wittgenstein's Tractatus Logico-Philosophicus.* Leibniz Company; also available as ebook, go to `https://archive.org/index.php` and search for "Huelster Introduction to Wittgenstein's Tractatus".

[Hyman, 1982] Hyman, A. (1982). *Charles Babbage: Pioneer of the Computer.* Princeton University Press.

[Kahneman, 2011] Kahneman, D. (2011). *Thinking, Fast and Slow.* Farrar, Straus, and Giroux.

[Kiss, 1999] Kiss, E. (1999). *Mathematical Gems from the Bolyai Chests.* Akadémiai Kiadó, Budapest; Typotex LTD, Budapest.

[Königliche Gesellschaft der Wissenschaften, Göttingen, 1900] Königliche Gesellschaft der Wissenschaften, Göttingen (1900). *Carl Friedrich Gauss Werke.* Teubner Verlag; available as ebook at

http://gdz.sub.uni-goettingen.de/dms/load/img/?PPN=PPN2
36010751&IDDOC=136917.

[Lakatos, 2015] Lakatos, I. (2015). *Proofs and Refutations*. Cambridge University Press, reissue edition.

[Lakoff and Núñez, 2000] Lakoff, G. and Núñez, R. E. (2000). *Where Mathematics Comes From*. Basic Books.

[Lenzen, 2004] Lenzen, W. (2004). *Calculus Universalis: Studien zur Logic von G. W. Leibniz*. Mentis Verlag.

[Livio, 2005] Livio, M. (2005). *The Equation That Couldn't Be Solved*. Simon & Schuster.

[Møller, 2014] Møller, A. R. (2014). *Neuroplasticity and its Dark Sides*. Aage R. Møller Publishing.

[Møller, 2015] Møller, A. R. (2015). New Developments in Neuroscience. *Journal of Integrated Creative Studies*, vol. 2015.

[Monk, 1990] Monk, R. (1990). *Ludwig Wittgenstein: The Duty of Genius*. Penguin Books.

[Morgan, 1847] Morgan, A. D. (1847). *Formal Logic: or The Calculus of Inference, Necessary and Probably*. Taylor and Walton; go to https://archive.org/index.php and search for "De Morgan Formal Logic".

[Moseley et al., 2016] Moseley, R., Kiefer, M., and Pulvermüller, F. (2016). Grounding and Embodiment of Concepts and Meaning - A neurological perspective. In *Foundations of Embodied Cognition, Vol 1: Perceptual and Emotional Embodiment*, Y. Coello and M. H. Fischer, eds. Routledge.

[Netz and Noel, 2007] Netz, R. and Noel, W. (2007). *The Archimedes Codex*. Weidenfeld & Nicolson, paperback Phoenix.

[Newman, 1956] Newman, J. R. (1956). *The World of Mathematics, Vols. I-IV*. Simon & Schuster; go to https://archive.org/index.php and search for "james newman world of mathematics".

[Ossendrijver, 2016] Ossendrijver, M. (2016). Ancient Babylonian astronomers calculated Jupiter's position from the area under a time-velocity graph. *Science*, vol. 351.

[Peckhaus, 1997] Peckhaus, V. (1997). *Mathesis universalis und allgemeine Wissenschaft. Leibniz und die Wiederentdeckung der formalen Logik im 19. Jahrhundert.* Akademie-Verlag (Logica Nova).

[Penrose, 2004] Penrose, R. (2004). *The Road to Reality.* Jonathan Cape, London.

[Polkinghorne, 2011] Polkinghorne, J., editor (2011). *Meaning in Mathematics.* Oxford University Press.

[Pulvermüller et al., 2013] Pulvermüller, F., Cappelle, B., and Shtyrov, Y. (2013). Brain basis of meaning, words, constructions, and grammar. In *The Oxford Handbook of Construction Grammar*, T. Hoffmann and G. Trousdale, eds. Oxford Universitiy Press.

[Rudman, 2007] Rudman, P. S. (2007). *How Mathematics Happened.* Prometheus Books.

[Russell and Norvig, 2010] Russell, S. and Norvig, P. (2010). *Artificial Intelligence - A Modern Approach.* Third edition. Prentice Hall.

[Sautoy, 2003] Sautoy, M. d. (2003). *The Music of the Primes: Searching to Solve the Greatest Mystery in Mathematics.* HarperCollins.

[Schoenflies, 1927] Schoenflies, A. (1927). Die Krisis in Cantor's mathematischem Schaffen. *Acta Mathematica*, vol. 50.

[Singh, 1997] Singh, S. (1997). *Fermat's Enigma: The Epic Quest to Solve the World's Greatest Mathematical Problem.* Walker.

[Sørnes, 2008] Sørnes, T. (2008). *The Clockmaker Rasmus Sørnes.* Borgarsyssed Museum, Sarpsborg, Norway.

[Stifel, 1544] Stifel, M. (1544). *Arithmetica Integra.* Johannes Petreius, Nürnberg.

[Théra, 2017] Théra, M. (2017). Homo sapiens, homo ludens. *Journal of Optimization Theory and Applications*, vol. 172.

[Thiel, 1982] Thiel, C. (1982). *Erkenntnistheoretische Grundlagen der Mathematik.* Gerstenberg Verlag.

[Waldvogel, 2012] Waldvogel, J. (2012). Jost Bürgi and the discovery of the logarithms. Research Report No. 2012-43, Eidgenössische Technische Hochschule (Swiss Federal Institute of Technology), Zürich.

[Walsdorf et al., 2014] Walsdorf, A., Badur, K., Stein, E., and Kopp, F. O. (2014). *Das letzte Original, Die Leibniz-Rechenmaschine der Gottfried Wilhelm Leibniz Bibliothek.* Gottfried Wilhelm Leibniz Bibliothek, Hannover.

[Weyl, 1921] Weyl, H. (1921). Über die neue Grundlagenkrise der Mathematik. *Mathematische Zeitschrift*, vol. 10.

[Whitehead, 1978] Whitehead, A. N. (1978). *Process and Reality: An Essay in Cosmology.* The Free Press.

[Wigner, 1960] Wigner, E. P. (1960). The unreasonable effectiveness of mathematics in the natural sciences. *Communications on Pure and Applied Mathematics*, vol. 13.

[Wilson, 1828] Wilson, J. (1828). *The Tetrabiblos; or, Quadripartite of Ptolemy.* William Hughes; see Google books "wilson the tetrabiblos".

[Wittgenstein, 1956] Wittgenstein, L. (1956). *Remarks on the Foundation of Mathematics.* Basil Blackwell; paperback MIT Press.

[Wittgenstein, 1958] Wittgenstein, L. (1958). *Philosophical Investigations.* Basil Blackwell; available at https://drive.google.com/file/d/0Bw-duXxYihdvWVlFaUhzclY5Vmc/edit.

[Wittgenstein, 1963] Wittgenstein, L. (1963). *Tractatus Logico-Philosophicus.* Routledge & Kegan Paul Ltd; go to people.umass.edu/klement/tlp/tlp.pdf for the German version and two translations into English.

[Wittgenstein, 1978] Wittgenstein, L. (1978). *Remarks on Colour.* University of California Press.

[Wolchover, 2013] Wolchover, N. (2013). Dispute over Infinity Divides Mathematicians. *Quanta Magazine*, December, 2013.

[Zacks, 2015] Zacks, J. (2015). *Flicker: Your Brain on Movies.* Oxford University Press.

[Zeki et al., 2014] Zeki, S., Romaya, J. P., Benincasa, D. M. T., and
  Atiyah, M. F. (2014). The experience of mathematical beauty and
  its neural correlates. *Frontiers in Human Neuroscience*, vol. 8.

[Zuse, 1993] Zuse, K. (1993).   *Der Computer–Mein Lebenswerk*.
  Springer-Verlag.

Acknowledgements

The writing of this book would have been impossible without the assistance of a large number of persons and institutions.

In particular, the following persons reviewed part or all of the material: D. Abbott, A. Alexander, J. Barrer, R. Chandrasekaran, T. Dénes, D. Everett, P. Gritzmann, M. Grötschel, R. Hersh, A. Møller, M. Opperud, M. Ossendrijver, F. Staudacher, I. H. Sudborough, L. Sullivan, I. Truemper, J. Truemper, S. Truemper, U. Truemper, J. Waldvogel, A. Yousefpour, and G. Ziegler.

The following persons and institutions provided information, contributed material, and generally helped in the creation of the book: D. Abbott, Academy of Sciences Berlin-Brandenburg, I. Adler, Arithmeum Bonn, Biblioteka IM PAN Warsaw, A. Borutzky, W. A. Casselman, Cohen family, A. Czader, D. van Dalen, T. Dénes, V. Enke, D. Everett, D. K. Frasier, V. Gallese, Gottfried Wilhelm Leibniz Bibliothek Hannover, I. Grötschel, M. Grötschel, G. Gupta, R. Hersh, K. Hamlen, Harvard University, M. Hittinger, D. S. Hochbaum, Institute for Advanced Study Princeton, S. Janeczko, D. Kahneman, H. Khelif, E. Knobloch, J. Koblitz, B. Korte, W. Lenzen, Lilly Library of University of Indiana, P. Mazaika, J. Miller, A. Møller, A. Nilsson, M. Ossendrijver, A. Ostrovsky, V. Peckhaus, S. Pitkin, Polish Academy of Sciences Warsaw, F. Pulvermüller, RAND Corporation, T. Reggev, G. Rinaldi, B. S. Shylaja, M. Skutella, T. Sørnes, F. Staudacher, Stanford University, I. Truemper, University of Texas at Dallas, J. Waldvogel, B. Wildenthal, A. Yousefpour, J. M. Zacks, and H. Zuse.

We very much thank all these persons and institutions for their help.

K. T.

Index

www.ingramcontent.com/pod-product-compliance
Lightning Source LLC
Chambersburg PA
CBHW030912090426
42737CB00007B/164